Contents

Preface

The use of the first person singular to refer to the author of this work should not be taken to imply that 'I' am the originator of the perspectives developed here. I must, of course, take sole responsibility for the particular formulations employed, and for the errors of fact and argument which no doubt remain in the text. However, any credit for this work is due not so much to me as to the wider circle of those who have been involved in planning, organising and discussing this project over the last five years. Members of the media group of the Centre for Contemporary Cultural Studies at the University of Birmingham have helped in various ways, but the project would not have been possible without the involvement of Charlotte Brunsdon, Ian Connell and Stuart Hall. I owe them an enormous debt.

D.M.

1 Audience Research: The Traditional Paradigms

It is not my purpose to provide an exhaustive account of mainstream sociological research in mass communications. I do, however, offer a resumé of the main trends and of the different emphases within that broad research strategy, essentially for two reasons: first because this project is framed by a theoretical perspective which represents, at many points, a different research paradigm from that which has dominated the field to date. Second, because there are points where this approach connects with certain important 'breaks' in that previous body of work, or else attempts to develop, in a different theoretical framework, lines of inquiry which mainstream research opened up but did not follow through.

Mainstream research can be said to have been dominated by one basic conceptual paradigm, constructed in response to the 'pessimistic mass society thesis' elaborated by the Frankfurt School. That thesis reflected the breakdown of modern German society into fascism, a breakdown which was attributed, in part, to the loosening of traditional ties and structures and seen as leaving people atomised and exposed to external influences, and especially to the pressure of the mass propaganda of powerful leaders, the most effective agency of which was the mass media. This 'pessimistic mass society thesis' stressed the conservative and reconciliatory role of 'mass culture' for the audience. Mass culture suppressed 'potentialities' and denied awareness of contradictions in a 'one-dimensional world'; only art, in fictional and dramatic form, could preserve the qualities of negation and transcendence.

Implicit here was a 'hypodermic' model of the media which were seen as having the power to 'inject' a repressive ideology directly into the consciousness of the masses. Katz and Lazarsfeld (1955), writing of this thesis, noted that:

> 'The image of the mass communication process entertained by researchers had been, firstly, one of "an atomistic mass" of millions of readers, listeners and movie-goers, prepared to receive the message; and secondly . . . every Message [was conceived of] as a direct and powerful stimulus to action which would elicit immediate response.' (p.16)

The emigration of the leading members of the Frankfurt School (Adorno, Marcuse, Horkheimer) to America during the 1930s led to the development of a specifically 'American' school of research in the forties and fifties. The Frankfurt School's 'pessimistic' thesis, of the link between 'mass society' and fascism, and the role of the media in cementing it, proved unacceptable to American researchers. The 'pessimistic' thesis proposed, they argued, too direct and unmediated an impact by the media on its audiences; it took too far the thesis that all intermediary social structures between leaders/media and the 'masses' had broken down; it didn't accurately reflect the 'pluralistic' nature of American society; it was – to put it shortly – sociologically naive. Clearly, the media had social effects; these must be examined, researched. But, equally clearly, these 'effects' were neither all-powerful, simple, nor even direct. The nature of this complexity and indirectness too had to be demonstrated and researched. Thus, in reaction to the Frankfurt School's predilection for 'critical' social theory and qualitative and philosophical analysis, the American researchers developed what began as a quantitative and positivist methodology for empirical radio audience research into the 'Sociology of Mass Persuasion'.

It must be noted that both the 'optimistic' and 'pessimistic' paradigms embodied a shared implicit theory of the dimensions of power and 'influence' through which the powerful (leaders and communicators) were connected to the powerless (ordinary people, audiences). Broadly speaking, operating within this paradigm, the different styles and strategies of research may then be characterised as a series of oscillations between two different, sometimes opposed, points in this 'chain' of communication and command. On the one hand, message-based studies, which moved from an analysis of the content of messages to their 'effects' on audiences; and, on the other, audience-based studies, which focused on the social characteristics, environment and, subsequently, 'needs' which audiences derived from, or brought to, the 'message'.

Many of the most characteristic developments within this paradigm have consisted either of refinements in the way in which the message-effect link has been conceptualised and studied; or of developments in the ways in which the audience and its needs have been examined. Research following the first strategy (message-effects) has been, until recently, predominantly behaviourist in general orientation: how the behaviour of audiences reflects the influences on them of the messages they receive. When a concern with 'cognitive' factors was introduced into the research it modified, without replacing, this behavioural orientation: messages could be seen to have effects only if a change of mind was followed by a change in behaviour (e.g. advertising campaigns leading to a change in commodity-choices). Research of the second type (audience-based) has been largely structural-functional in orientation, focusing on the social characteristics of different audiences, reflecting their different degrees of 'openness' to the messages they received. When a 'cognitive' element was introduced here, it modified without replacing this functional perspective: differences in audience response were related to differences in individual needs and 'uses'.

We will look in a moment at the diverse strategies through which this basic conceptual paradigm was developed in mainstream research. It is not until recently that a conceptual break with this paradigm has been mounted

Ref. only.

BFI 1000

X

ME 2✓

BFI TELEVISION MONOGRAPH

11

The *Nationwide* Audience: Structure and Decoding

David Morley

British Film Institute
127 Charing Cross Road, London WC2H 0EA
1980

Television Monograph No. 10, *Everyday Television: 'Nationwide'*, was a 'reading' of the programme which attempted to reveal the ways in which *Nationwide* constructs for itself an image of its audience. The present volume is an extension of that study, this time showing through detailed empirical work how different audiences actually 'decoded' some *Nationwide* programmes. The analysis is set in the context of a substantial critique of previous audience research.

The author

David Morley is Lecturer in Communications at Lanchester Polytechnic.

ISBN 0 85170 097 7

in the research field, one which has attempted to grasp communication in terms neither of societal functions nor of behavioural effects, but in terms of social meanings. This latter work is described here as the 'interpretative' as against the more dominant 'normative' paradigm, and it does constitute a significant break with the traditional mainstream approach. The approach adopted in this project shares more with the 'interpretative' than with the traditional paradigm, but I wish to offer a critique, and to propose a departure from, *both* the 'normative' and 'interpretative' paradigms, as currently practised.

The 'Normative' Paradigm

Post-war American mass communications research made a three-dimensional critique of the 'pessimistic' mass society thesis: refuting the arguments that informal communication played only a minor role in modern society, that the audience was a 'mass' in the simple sense of an aggregation of socially-atomised individuals, and that it was possible to equate directly content and effect.

In an early work which was conceptually highly sophisticated, Robert Merton (1946) first advanced this challenge with his case study (*Mass Persuasion*) of the Kate Smith war bond broadcasts in America. Though this work was occasionally referred to in later programmatic reviews of the field, the seminal leads it offered have never been fully followed through. Merton argued that research had previously been concerned almost wholly with the 'content rather than the effects of propaganda'. Merton granted that this work had delivered much that had been of use, in so far as it had focused on the 'appeals and rhetorical devices, the stereotypes and emotive language which made up the propaganda material'. But the 'actual processes of persuasion' had gone unexamined, and as a consequence 'the effect' of the materials studied had typically been assumed or inferred, particularly by those who were concerned with the malevolent effect of 'violent' content. Merton challenged this exclusive reliance on inference from content to predicted effects.

This early work of Merton is singular in several respects, not least for the attempt it made to connect together the analysis of the message with the analysis of its effects. Social psychology had pointed to 'trigger phrases which suggest to us values we desire to realise'. But, Merton asked, 'Which trigger phrases prove persuasive and which do not? Further, which people are persuaded and which are not? And what are the processes involved in such persuasion and in resistance to persuasive arguments?' To answer these questions Merton correctly argued that we had to 'analyse both the content of propaganda and the responses of the audience. The analysis of content . . . gives us clues to what might be effective in it. The analysis of responses to it enables us to check those clues.' Merton thus retained the notion that the message played a determining role for the character of the responses that were recorded, but argued against the notion that this was the only determination and that it connected to response in a simple cause and effect relationship; indeed, he insisted that the message 'cannot adequately be interpreted if it is severed from the cultural context in which it occurred.'

Merton's criticisms did not lead to any widespread reforms in the way in which messages were analysed as such. Instead, by a kind of reversal, it opened the road to an almost exclusive pre-occupation with receivers and reception situations. The emphasis shifted to the consideration of small groups and opinion leaders, an emphasis first developed in Merton's own work on 'influentials' and 'reference groups', and later by Katz and Lazarsfeld. Like Merton, they rejected the notion that influence flowed directly from the media to the individual; indeed, in *Personal Influence* (1955) they developed the notions of a 'two-step flow of communication' and of the importance of 'opinion leaders' within the framework of implications raised by small group research. From several studies in this area it had become obvious, according to Katz and Lazarsfeld, that 'the influence of mass media are not only paralleled by the influence of people . . . [but also] . . . refracted by the personal environment of the ultimate consumer.'

The 'hypodermic model' – of the straight, unmediated effect of the message – was decisively rejected in the wake of this 'rediscovery of the primary group' and its role in determining the individual's response to communication. In *The People's Choice* it was argued that there was little evidence of people changing their political behaviour as a result of the influence of the media: the group was seen to form a 'protective screen' around the individual. This was the background against which Klapper (1960) summed up: 'persuasive communications function far more frequently as an agent of reinforcement than as an agent of change . . . reinforcement, or at least constancy of opinion, is typically found to be the dominant effect . . .'

From 'Effects' to 'Functions' . . . and back again

The work outlined above, especially that of Merton, marks a watershed in the field. I have discussed it in some detail because, though there have been many subsequent initiatives in the field, they have largely neglected the possible points of development which this early work touched on.

The intervening period is, in many ways, both more dismal and less fruitful for our purposes. The analysis of content became more quantitative, in the effort to tailor the description of vast amounts of 'message material' for the purposes of effects analysis. The dominant conception of the message here was that of a simple 'manifest' message, conceived on the model of the presidential or advertising campaign, and the analysis of its content tended to be reduced, in Berelson's (1952) memorable phrase, to the 'quantitative description of the manifest content of communication'. The complexity of Merton's Kate Smith study had altogether disappeared.

Similarly, the study of 'effects' was made both more quantitative and more routine. In this climate Berelson and others predicted the end of the road for mass communications research.

A variety of new perspectives were suggested, but the more prominent were based on the 'social systems' approach and its cousin, 'functional analysis' (Riley and Riley, 1959), concerning themselves with the general functions of the media for the society as a whole (see Wright's (1960) attempt to draw up a 'functional inventory'). A different thread of the functionalist approach was more concerned with the subjective motives and interpretations of individual users. In this connection Katz (1959) argued that the approach crucially assumed that 'even the most potent of mass media content cannot ordinarily influence an individual who has no "use" for it in the social and psychological context in which he lives. The "uses" approach assumes that people's values, their interests . . . associations . . . social roles, are pre-potent, and that people selectively fashion what they see and hear.' This strand of the research work has of course, recently re-emerged in the work of the British 'uses and gratifications' approach, and has been hailed, after its long submergence, as the road forward for mass communications research.

These various functionalist approaches were promulgated as an alternative to the 'effects' orientation; nonetheless, a concern for effects remained, not least among media critics and the general public. This concern with the harmful effects or 'dysfunctions' of the media was developed in a spate of laboratory-based social-psychological studies which, in fact, followed this functionalist interlude. This, rather than the attempt to 'operationalise' either of the competing functionalist models, was the approach which dominated mass media research in the 1960s: the attempt to pin down, by way of stimulus-response, imitation and learning theory psychology approaches, applied under laboratory conditions, the small but quantifiable effects which had survived the 'optimistic' critique.

Bandura (1961) and Berkowitz (1962) were among the foremost exponents of this style of research, with their focus on the message as a simple, visual stimulus to imitation or 'acting out', and their attention to the consequences, in terms of violent behaviour and delinquency, of the individual's exposure to media portrayals of violence of 'filmed aggressive role models'. Halloran's study of television and violence in this country took its point of departure from this body of work.

During the mid to late sixties research on the effects of television portrayals of violence was revitalised and its focus altered, in the face of the student rebellion and rioting by blacks in the slum ghettoes of America (see *National Commission on Causes and Prevention of Violence: the Surgeon-General's Report*). Many of the researchers and representatives of the state who were involved in this work, in their concluding remarks, suggested that TV was not a principal cause of violence but rather, a contributing factor. They acknowledged, as did the authors of the *National Commission Report,* that 'television, of course, operates in social settings and its effects are undoubtedly mitigated by other social influences'. But despite this gesture to mitigating or intervening social influences, the conviction remained that a medium saturated with violence must have some direct effects. The problem was that researchers operating within the mainstream paradigm still could not form any decisive conclusions about the impact of the media. The intense controversy following the attempt of the Surgeon-General to quantify a 'measurable effect' of media violence on the public indicated how controversial and inconclusive the attempt to 'prove' direct behavioural effect remained.

Mick Counihan (1972), in a review of the field, summarised the development of mass communications research up to this point as follows:

> 'Once upon a time . . . worried commentators imputed an eternal omnipotence to the newly emerging media of mass communication. In the "Marxist" version of this model of their role, the media were seen as entirely manupulated by a shrewd ruling class in a bread and circuses strategy to transmit a corrupt culture and neofascist values – violence, dehumanised sex, consumer brainwashing, political passivity, etc. – to the masses. . .
>
> 'These instruments of persuasion on the one hand, and the atomised, homogenised, susceptible masses on the other, were conjoined in a simple stimulus-response model. However, as empirical research progressed, survey and experimental methods were used to measure the capacity of the media to change "attitude", "opinions", and "behaviour". In turn, the media-audience relationship was found to be not simple and direct, but complex and mediated. "Effects" could only be gauged by taking account of the "primary groups", opinion leaders and other factors intervening between the media and the audience member (hence the notion of "two-step" or "multiple-step" flow of information). . .
>
> 'Further emphasis shifted from "what the media do to people" to "what people do to the media", for audiences were found to "attend to" and "perceive" media messages in a selective way, to tend to ignore or to subtly interpret those messages hostile to their particular viewpoints. Far from possessing ominous persuasive and other anti-social power, the media were now found to have a more limited and, implicitly, more benign role in society: not changing, but "reinforcing" prior dispositions, not cultivating "escapism" or passivity, but capable of satisfying a great diversity of "uses and gratifications", not instruments of a levelling of culture, but of its democratisation.' (p. 43)

The Interpretative Paradigm

In the same period, a revised sociological perspective was beginning to make inroads on communications research. What had always been assumed was a shared and stable system of values among all the members of a society; this was precisely what the 'interpretative' paradigm put into question, by its assertion that the meaning of a particular action could not be taken for granted, but must be seen as problematic for the actors involved. Interaction was thus conceptualised as a process of interpretation and of 'mutual typification' by and of the actors involved in a given situation.

The advances made with the advent of this paradigm were to be found in its emphasis on the role of language and symbols, everyday communication, the interpretation of action, and an emphasis on the process of 'making sense' in interaction. However, the development of the interpretative paradigm in its ethno-methodological form (which turned the 'normative' paradigm on its head) revealed its weaknesses. Whereas the normative approach had focused individual actions exclusively as the reproduction of shared stable norms, the interpretative model, in its ethno-methodological form, conceived each interaction as the 'production' anew of reality. The problem here was often that although ethno-methodology could shed an interesting light on micro-processes of interpersonal communications, this was disconnected from any notion of institutional power or of structural relations of class and politics.

Aspects of the interactionist perspective were later taken over by the Centre for Mass Communications Research at Leicester University, and the terms in which its Director, James Halloran (1970), discussed the social effects of television gave some idea of its distance from the normative paradigm; he spoke of the 'trend away from . . . the emphasis on the viewer as tabula-rasa . . . just waiting to soak up all that is beamed at him. Now we think in terms of interaction or exchange between the medium and audience, and it is recognised that the viewer approaches every viewing situation with a complicated piece of filtering equipment.'

This article also underlined the need to take account of 'subjective definitions of situations and events', without going over fully to the 'uses and gratifications' position. Halloran recast the problematic of the 'effects of television' in terms of 'pictures of the world, the definitions of the situation and problems, the explanations, alternative remedies and solutions which are made available to us . . .' The empirical work of the Leicester Centre at this time marked an important shift in research from forms of behavioural analysis to forms of cognitive analysis. *Demonstrations and Communications* (1970b) attempted to develop an analysis of 'the communication process as a whole', studying 'the production process, presentation and media content as well as the reactions of the viewing and reading public'. This latter aspect of the research was further developed by Elliott in his study *The Making of a Television Series* (1972), especially the notion of public communication as a circuit relaying messages from 'the society as source' to 'the society as audience'.

The developing emphasis on the levels of cognitive and interpretative analysis made it possible for the whole question of the 'influence of the media' to be reopened and developed. As Hartmann and Husband noted (1972), part of the problem with the theories of 'media impotence' was that their inadequate methodological orientations predisposed them to this conclusion, for:

> 'to look for effects in terms of simple changes of attitude may be to look in the wrong place . . . part of the high incidence of null results in attempts to demonstrate the effects of mass communications lies in the nature of the research questions asked . . .
>
> . . . it may be that the media have little immediate impact on attitudes as commonly assessed by social scientists but it seems likely that they have other important effects. In particular they would seem to play a major part in defining for people what the important issues are and the terms in which they should be discussed.'

This perspective allows for a re-evaluation of 'effects' studies such, for instance, as that of Nordenstreng (1972). Nordenstreng has argued, on the basis of his research on Finnish TV audiences, that watching TV news is a 'mere ritual' for the audience which has 'no effect'. His research showed that, while 80 per cent of Finns watched at least one news broadcast per day, when interviewed the next day they could remember hardly anything of the specific information given by the news; the main impression retained was that 'nothing much had happened'. On this basis Nordenstreng argued that the 'content of the news is indifferent to them'.

Hartmann and Husband's argument leads beyond this difficulty in so far as it clarifies the point that one cannot argue that just because an audience cannot remember specific content – names of ministers, etc. – that therefore a news broadcast has had 'no effect'. The interpretative paradigm led to a focus on the level of 'definitions' and 'agendas' of issues, rather than on specific content items. The important point is that while an audience may retain little, in terms of specific information, they may well retain general 'definitions of the order of things', ideological categories embedded in the structure of the specific content. Indeed Hartmann and Husband's research on race and the media precisely focused the impact of the media on definitional frameworks, rather than on specific attitudes or levels of information.

However, despite these important conceptual advances, the Centre's work has been marred by an inadequate conception of the message. Following Gerbner's work, this was conceived of as the 'coin of communicative exchanges', but at the same time, it was thought to have only one 'real meaning'. This notion is also evident in the more recent audience research project *Understanding TV* (O. Linné and K. Marossi; Danish Radio Research Report, 1976) on which Halloran collaborated. Here again the principal concern of the research was to check whether the real/intended message of the communicators had 'got through' to the audience. In short, the framework was effectively that of a 'comprehension' study – as to whether the audience had 'understood' the (pre-defined) meaning of the programme (a documentary on Vietnam) – rather than one concerned with the various possible kinds of meanings and understandings that were generated from the programme by the different sections of its audience. The implications of this misconception of the message can best be assessed by situating it in relation to the overall model of the communication process proposed here.

The Communication Circuit: Breaks and Oppositions

Much of the work of the Centre for Contemporary Cultural Studies media group over the last few years represents an attempt to fill out and elaborate a model of the 'communication process as a whole' of the kind called for by Halloran (1970a). Necessarily, this has involved detailed work on particular moments in that process, or on the links in the communication chain between such 'moments'. This project was intended to refine certain hypotheses about one such privileged moment: that which links the production of the TV message to its reception by certain kinds of audiences. It is not necessary to spell this general model out in detail here, but it may be worthwhile to specify some general theoretical outlines about how such a model must be 'thought'. Though conceived as a communication chain or 'flow', the TV discourse is *not* a flow between equivalent elements. Thus the chain must be conceived as one which exhibits breaks and discontinuities. It connects 'elements' or moments which are not identical, and which have a different position in the hierarchy of communication and an identifiable structure of their own. Hence, the flow between one point and another in the communication chain depends *a*) on two points (sender and receiver) which have their own internal structure (the structure of TV production, the structures of reception); which *b*) being different, and yet linked, therefore require *c*) specific mechanisms or forms to articulate them into a unity. That unity is necessarily *d*) a complex not a simple unity; that is, *e*) one where the articulations, though they will *tend* to flow in a certain way, and to exhibit a certain logic, are neither closed nor finally determined (cf. Hall, 1973).

In the first place this is to reject those theories which 'collapse' the specificity of the media's mode of operation by picturing the media as totally controlled and manipulated by the state (Althusser, 1971; Garnham, 1973), or by arguing directly from the economic structure of media institutions (concentration of ownership) to the 'effects' of this structure at the level of power and ideology (cf. Murdock & Golding, 1973 *et seq.*) – the production of cultural artefacts which legitimate the consensus – without giving specific weight in the analysis to the process of symbolisation itself.

However, to take account of these 'breaks' in the communications chain is not, in itself, enough; we must allow for the break at the reception end of the chain, and *not* presume that audiences are bound to programming in a transparent relationship, in which all messages are read 'straight'. That the broadcaster will attempt to establish this relationship of 'complicity' with the audience is undoubtedly true, (as argued in *'Everyday Television: 'Nationwide'*, for instance, via the provision of 'points of identification' within the programme) but to assume the success of these attempts is to assume away the crucial problem.

The Message: Encoding and Decoding

The premises on which this approach is based can be outlined as follows: 1) The production of a meaningful message in the TV discourse is always problematic 'work'. The same event can be encoded in more than one way. The study here is, then, of how and why certain production practices and structures tend to produce certain messages, which embody their meanings in certain recurring forms. 2) The message in social communication is always complex in structure and form. It always contains more than one potential 'reading'. Messages propose and prefer certain readings over others, but they can never become wholly closed around one reading. They remain polysemic. 3) The activity of 'getting meaning' from the message is also a problematic practice, however transparent and 'natural' it may seem. Messages encoded one way can always be read in a different way.

In this approach, then, the message is treated neither as a unilateral sign, without ideological 'flux' (cf. Woolfson, 1976), nor, as in 'uses and gratifications', as a disparate sign which can be read any way, according to the need-gratification structure of the decoder. The TV message is treated as a complex sign, in which a preferred reading has been inscribed, but which retains the potential, if decoded in a manner different from the way in which it has been encoded, of communicating a different meaning. The message is thus a structured polysemy. It is central to the argument that all meanings do not exist 'equally' in the message: it has been structured in dominance, although its meaning can never be totally fixed or 'closed'. Further, the 'preferred reading' is itself part of the message, and can be identified within its linguistic and communicative structure.

Thus, when analysis shifts to the 'moment' of the encoded message itself, the communicative form and structure can be analysed in terms of what is the preferred reading; what are the mechanisms which prefer one, dominant reading over the other readings; what are the means which the encoder uses to try to 'win the assent of the audience' to his/her preferred reading of the message? Special attention can be given here to the control exercised over meaning and to the 'points of identification' within the message which transmit the preferred reading to the audience.

The presentational or interviewing styles employed will also be critical for winning the assent of the audience to the encoder's reading of the message. Here we can attempt to reintegrate into the analysis some of the insights relating to the congruence between the internal structure of the message and the objective form of the audience's understanding and responses which were, in many ways, first empirically pursued in Merton's early study discussed above. Krishan Kumar (1977) has pointed to the key role of the 'professional broadcasters', the linkmen and presenters who from the public's point of view 'constitute the BBC'; as Kumar argues, these men 'map out for the public the points of identification with the BBC'; they are the 'familiar broadcasting figures with whom the audience can identify'.

It is precisely the aim of the presenter to achieve this kind of audience-identification. The point is that it is through these identification mechanisms, I would suggest, in so far as they do gain the audience's 'complicity', that the preferred readings are 'suggested' to the audience. It is when these identificatory mechanisms are attenuated or broken that the message will be decoded in a different framework of meaning from that in which it was encoded.

We can now turn to the moment in the communicative chain where these encoded messages are received. Here I would want to insist that audiences, like the producers of messages, must also undertake a specific kind of 'work' in order to read meaningfully what is transmitted. Before messages can have 'effects' on audiences, they must be decoded. 'Effects' is thus a shorthand, and inadequate, way of marking the point where audiences differentially read and make sense of messages which have been transmitted, and act on those meanings within the context of the rest of their situation and experience. Moreover, we must assume that there will be no necessary 'fit' or transparency between the encoding and decoding ends of the communication chain (cf. Hall, 1973). It is precisely this lack of transparency and its consequences for communication which this project has attempted to investigate. It is the problem of conceptualising the media audience to which we must now turn.

Undoubtedly, the dominant perspective in audience research, certainly in this country, in recent years has been the 'uses and gratifications' perspective developed by Blumler *et al.* It may therefore be best to attempt to define the perspective employed in this project by way of a critique of the limitations of the 'uses and gratifications' problematic.

2 'What people do with the media': Uses, Gratifications, Meanings

The realisation within mass media research that one cannot approach the problem of the 'effects' of the media on the audience as if contents impinged directly onto passive minds; that people in fact assimilate, select from and reject communications from the media, led to the development of the 'uses and gratifications' model; Halloran advising us that:

> 'We must get away from the habit of thinking in terms of what the media do to people and substitute for it the idea of what people do with the media.'

This approach highlighted the important fact that different members of the mass media audience may use and interpret any particular programme in a quite different way from how the communicator intended it, and in quite different ways from other members of the audience. Rightly it stressed the role of the audience in the construction of meaning.

However, this 'uses and gratifications' model suffers from fundamental defects in at least two respects.

1) As Hall (1973) argues, in terms of its overestimation of the 'openness' of the message:

> 'Polysemy must not be confused with pluralism. Connotative codes are not equal among themselves. Any society/culture tends, with varying degrees of closure, to impose its segmentations . . . its classifications of the . . . world upon its members. There remains a dominant cultural order, though it is neither univocal or uncontested.'

While messages can sustain, potentially, more than one reading, and:

> '. . . there can be no law to ensure that the receiver will "take" the preferred or dominant reading of an episode . . . in precisely the way in which it has been encoded by the producer.' (op. cit., p. 13)

yet still the message is 'structured in dominance' by the preferred reading. The moment of 'encoding' thus exerts, from the production end, an 'over-determining' effect (though not a fully determined closure) on the succeeding moments in the communicative chain.

As Elliott (1973) rightly argues, one fundamental flaw in the 'uses and gratifications' approach is that its implicit model of the communication process fails to take into account the fact that television consumption is:

> 'more a matter of availability than of selection . . . [In this sense] availability depends on familiarity . . . The audience has easier access to familiar genres partly because they understand the language and conventions and also because they already know the social meaning of this type of output with some certainty.' (p. 21)

Similarly Downing (1974) has pointed to the limitations of the assumption (built into the 'uses and gratifications' perspective) of an unstructured mass of 'differential interpretations' of media messages. As he points out, while in principle a given 'content' may be interpreted by the audience in a variety of ways:

> 'In practice a very few of these views will be distributed throughout the vast majority of the population, with the remainder to be found only in a small minority. [For] given a set of cultural norms and values which are very dominant in the society as a whole (say the general undesirability of strikes) and given certain stereotypes (say that workers and/or unions initiate strikes) only a very sustained and carefully argued and documented presentation of any given strike is likely to challenge these values and norms.' (p. 111)

2) The second limitation of the 'uses and gratifications' perspective lies in its insufficiently sociological nature. Uses and gratifications is an essentially psychologistic problematic, relying as it does on mental states, needs and processes abstracted from the social situation of the individuals concerned – and in this sense the 'modern' uses and gratifications approach is less 'sociological' than earlier attempts to apply this framework in the USA. The earlier studies dealt with specific types of content and specific audiences, whereas 'modern' uses and gratifications tends to look for underlying structures of need and gratification of psychological origin, without effectively situating these within any socio-historical framework.

As Elliott (1973) argues, the 'intra-individual' processes with which uses and gratifications research deals:

> 'can be generalised to aggregates of individuals, but they cannot be converted in any meaningful way into social structure and process . . .' (p. 6)

because the audience is still here conceived of as an atomised mass of individuals (just as in the earlier 'stimulus-response' model) abstracted from the groups and subcultures which provide a framework of meaning for their activities.

Graham Murdock (1974) rightly argues that:

> 'In order to provide anything like a satisfactory account of the relationship between people's mass media involvements and their overall social situation and meaning system, it is necessary to start from the social

setting rather than the individual: to replace the idea of personal "needs" with the notion of structural contradiction; and to introduce the notion of subculture . . . Subcultures are the meaning systems and modes of expression developed by groups in particular parts of the social structure in the course of their collective attempt to come to terms with the contradictions in their shared social situation . . . They therefore provide a pool of available symbolic resources which particular individuals or groups can draw on in their attempt to make sense of their own specific situation and construct a viable identity.' (p. 213)

This is to argue for the essentially social nature of consciousness as it is formed through language much in the way that Voloshinov does (quoted in Woolfson, 1976):

> 'Signs emerge after all, only in the process of interaction between one individual consciousness and another. And the individual consciousness itself is filled with signs. Consciousness becomes consciousness only once it has been filled with ideological (semiotic) content, consequently only in the process of social interaction.' (p. 168)

As Woolfson remarks of this, the sign is here seen as vehicle of social communication, and as permeating the individual consciousness, so that consciousness is seen as a socio-ideological fact. From this position Woolfson argues that:

> 'speech utterances are entirely sociological in nature. The utterance is always in some degree a response to something else. It is a product of inter-relationship and its centre of gravity therefore lies outside the individual speaker him/herself.' (p. 172)

Thus utterances are to be examined not as individual, idiosyncratic expressions of a psychological kind, but as sociologically regulated both by the immediate social situation and by the surrounding socio-historical context; utterances form a:

> 'ceaseless stream of dialogic inter-change [which is the] generative process of a given social collective.' (Woolfson, op.cit.)

What Woolfson argues here in relation to the need to redefine the analysis of 'individual' speech utterances – as the analysis of the communicative utterances of 'social individuals' – I would argue in relation to the analysis of 'individual' viewing patterns and responses. We need to break fundamentally with the 'uses and gratifications' approach, its psychologistic problematic and its emphasis on individual differences of interpretation. Of course, there will always be individual, private readings, but we need to investigate the extent to which these individual readings are patterned into cultural structures and clusters. What is needed here is an approach which links differential interpretations back to the socio-economic structure of society, showing how members of different groups and classes, sharing different 'culture codes', will interpret a given message differently, not just at the personal, idiosyncratic level, but in a way 'systematically related' to their socio-economic position. In short we need to see how the different sub-cultural structures and formations within the audience, and the sharing of different cultural codes and competencies amongst different groups and classes, 'determine' the decoding of the message for different sections of the audience.

Halloran (1975) has argued that the:

> 'real task for the mass communications researcher is . . . to identify and map out the different sub-cultures and ascertain the significance of the various sub-codes in selected areas governed by specific broadcasting or cultural policies.'

This is necessary, Halloran argues, because we must see that:

> 'the TV message . . . is not so much a message . . . [but] more like a message-vehicle containing several messages which take on meanings in terms of available codes or subcodes. We need to know the potential of each vehicle with regard to all the relevant sub-cultures.' (p. 6)

This is to propose a model of the audience, not as an atomised mass of individuals, but as a number of sub-cultural formations or groupings of 'members' who will, as members of those groups, share a cultural orientation towards decoding messages in particular ways. The audience must be conceived of as composed of clusters of socially situated individual readers, whose individual readings will be framed by shared cultural formations and practices pre-existent to the individual: shared 'orientations' which will in turn be determined by factors derived from the objective position of the individual reader in the class structure. These objective factors must be seen as setting parameters to individual experience, although not 'determining' consciousness in a mechanistic way; people understand their situation and react to it through the level of subcultures and meaning systems.

This brings us, in the first instance, to the problem of the relationship between social structure and ideology. The work of Basil Bernstein (1971) and others in the field of educational sociology is of obvious relevance here and some extrapolations on the possible significance of that work for media audience research are made in Morley (1975), 'Reconceptualising the Audience'. Rather than rehearse that argument in detail I will attempt in the next chapter simply to outline the notorious problem of the relation of classes and codes and its significance for this audience research project.

3 Classes, Codes, Correspondences

Autonomy: Relative or Total?

In recent years one of the most significant interventions in the debate about the problem of 'determination', or the relation of class structure and ideology, has been that made by Paul Hirst and his associates. They have argued that the attempt to specify this determination is doomed to incoherence, on the grounds that either the determination must be total, in which case the specificity of the ideological or the 'level of signifying practices' is denied; or alternatively that the proper recognition of this autonomy precludes the specification of any such form of determination of the ideological. The argument is, of course, premised on the rejection of the concept of 'relative autonomy' derived from Althusser, and in particular on the rejection of the use of that concept within the field of cultural studies (cf. Coward, 1977).

John Ellis (1977) takes up this problem, basing his position on Hirst's arguments in *Economy and Society,* November 1976. He denies the sense of attempting to derive expectations as to ideological/political practices from class position, and denies the validity of any model of 'typical positions' (such as those embedded in the encoding/decoding model derived from Parkin). He argues that this is illegitimate, since: 'According to the conjuncture, shopkeepers, for example, can be voting communist, believing in collective endeavour.' ('The Institution of the Cinema', p. 58) So presumably, according to the 'conjuncture', shopkeepers can be decoding programmes in any number of different frameworks/codes, in an unstructured way.

This 'radical' formulation of the autonomy of signifying/ideological practices seems inadequate in two respects. Firstly, by denying the relevance of cultural contexts in providing for individuals in different positions in the social structure a differential range of options, the argument is reduced, by default, to a concern with the (random) actions (voting, decoding) of individuals abstracted from any socio-historical context except that of the (unspecified) conjuncture.

This return to methodological individualism must be rejected if we are to retain any sense of the audience as a structured complex of social collectivities of different kinds. It is the sense of Voloshinov's conception of the 'social individual' which must be retained if we are not to return to a conception of the individual who is (*contra* Marx) simply 'in society as an object is in a box'.

Secondly, and more importantly, the Ellis/Hirst approach simply seems to throw out the baby with the bathwater. The argument against a mechanistic interpretation in which:

> 'it is assumed that the census of employment category carries with it both political and ideological reflections' (Ellis op. cit., p. 65)

is of course, perfectly correct, precisely because this approach

> 'eliminates the need for real exploration of ideological representations in their specificity'

by assuming that members of category X hold beliefs of type Y 'as a function of their economic situation'.

However, there is no licence for moving from this position, as Ellis and Hirst do, to an argument that therefore *all* attempts to specify determination by class structure are misconceived. The argument here becomes polarised into an either/or, both poles of which are absurd: either total determination or total autonomy.

The problem is that shopkeepers do not act as Ellis hypothesises. The reason that bourgeois political science makes any kind of sense at all, even to itself, is precisely because it is exploring a structured field in which class determinations do, simply on a level of statistical probability, produce correlations and patterns. Now, simply to count these patterns may be a fairly banal exercise, but to deny their existence is ludicrous; the patterns are precisely what are to be explored, in their relation to class structures.

It is interesting to compare the Ellis/Hirst intervention with that of Harold Rosen (1972) in *Language and Class*, who begins to provide precisely the kind of non-mechanistic, non-economistic account of the determination of language by class and the action of language on class formation that Ellis and Hirst seem to consider impossible. Rosen attacks Basil Bernstein exactly for providing a mechanistic, economistic analysis. The working class is, in Bernstein's work, an undifferentiated whole, defined simply by economic position. Factors at the level of ideological and political practice which 'distinguish the language of Liverpool dockers from that of . . . Coventry car workers' (Rosen, p. 9) are ignored. However, Rosen precisely aims to extend the terms of the analysis by inserting these factors as determinate. He rejects the argument that linguistic code can be simply determined by 'common occupational function' and sees the need to differentiate within and across class categories in terms of ideological practice:

> 'history, traditions, job experience, ethnic origins, residential patterns, level of organisation . . .' (p. 6)

Yet the central concern remains. The intervention is called *Language and Class* and its force is produced directly by the attention paid (as against

Bernstein's mechanistic/economistic account) to the levels of ideological, discursive, and political practice. These factors are here inserted with that of class determination, and their extension into the field of decoding is long overdue – but the relative autonomy of signifying practices does not mean that decodings are not structured by class. *How* they are so structured, in what combinations for different sections of the audience, the relation of language, class and code, is 'a question which must be ethnographically investigated' (Giglioli, 1972, p. 10).

The question is really whether this is an irretrievably essentialist or mechanistic problematic. I would argue that a charge of mechanism cannot be substantiated. Indeed the formulation of structures, cultures and biographies (outlined in Critcher, 1978) clearly evades the polarity of either total determination or total autonomy, through the notion of structures setting parameters, determining the availability of cultural options and responses, not directly determining other levels and practices. This problematic, then, clearly is concerned with some form of determination of cultural competencies, codes and decodings by the class structure, while avoiding mechanistic notions.

The problem which this project was designed to explore was that of the extent to which decodings take place within the limits of the preferred (or dominant) manner in which the message has been initially encoded. However, the complementary aspect of this problem is that of the extent to which these interpretations or decodings are inflected by other codes and discourses which different sections of the audience inhabit. We are concerned here with the ways in which decoding is determined by the socially governed distribution of cultural codes between and across different sections of the audience: that is, the range of different decoding strategies and competencies in the audience.

To raise this as a problem for research is already to argue that the meaning produced by the encounter of text and subject cannot be 'read off' straight from textual characteristics; rather:

> 'what has to be identified is the use to which a particular text is put, its function within a particular conjucture, in particular institutional spaces, and in relation to particular audiences.' (Neale 1977, pp. 39–40)

The text cannot be considered in isolation from its historical conditions of production and consumption.

Thus the meaning of the text must be thought in terms of which set of discourses it encounters in any particular set of circumstances, and how this encounter may re-structure both the meaning of the text and the discourses which it meets. The meaning of the text will be constructed differently according to the discourses (knowledges, prejudices, resistances, etc.) brought to bear on the text by the reader and the crucial factor in the encounter of audience/subject and text will be the range of discourses at the disposal of the audience.

Here, of course, 'individuals do have different relations to sets of discourses in that their position in the social formation, their positioning in the real, will determine which sets of discourses a given subject is likely to encounter and in what ways it will do so.' (Willemen, 1978, pp. 66–7)

Clearly Willemen is here returning to the agenda a set of issues about the relation between social position and discursive formation which were at the core of the work in educational sociology generated by Bernstein's intervention, and developed in France by Bourdieu, Baudelot and Establet. Moreover, Willemen's work can be seen as a vital element in the development of such a theory. As he argues, determination is not to be conceived as a closed and final process:

> 'Having recognised the determining power of the real, it is equally necessary to recognise that the real is never in its place, to borrow a phrase from Lacan, in that it is always and only grasped as reality, that is to say, through discourse . . . the real determines to a large extent the encounter of/with discourses, while these encounters structure, produce reality, and consequently in their turn affect the subject's trajectory through the real.' (op. cit. pp. 67–8)

or as Neale (1977) would have it:

> 'audiences are determined economically, politically *and ideologically*.' [my emphasis] (p. 20)

Sociologisms

The emphasis given in the Neale quotation above is intended to highlight the chief problem with the other main body of theory which has attempted to deal with the question of the relation between ideology and social structure – or, in its own terms, between a dominant normative order, roles and individuals. This, of course, is that substantial body of work in sociology variously addressed as the problem of order, the problem of class consciousness, and so on.

The inadequacy of this work is that of its sociologism in so far as it attempts to immediately convert social categories (e.g. class) into meanings (e.g. ideological positions) without attention to the specific factors necessarily governing this conversion. Thus, it is of course inadequate to present demographic/sociological factors – age, sex, race, class position – as objective correlates or determinants of differential decoding positions without any attempt to specify *how* they intervene in the process of communication. The problem, I would suggest, is how to think the relative autonomy of signifying practices (which most of the sociological theory neglects) in combination with the operation of class/gender/race, etc., as determinations.

Those who have followed Hirst and others in their development of a theory of the autonomy and specificity of 'signification' repeatedly warn against the dangers of a reductionist form of analysis, which is concerned with the level of ideology and signification only for as long as it takes to discover which ideologies are being worn as the 'number plates' of which classes. We must take due note of Laclau's (1977) strictures against the 'metaphysical assignment to classes of certain ideological elements' (p. 99) and against a mode of analysis in which 'the discrimination of "elements" in terms of their class belonging and the abstract postulation of pure ideologies are mutually dependent aspects'. (p. 94) On this basis we can proceed to attempt to specify complex modes of determination which precisely start from the presupposition that 'ideological elements, taken in isolation, have no necessary class connotation, and that this connotation is only the result of the articulation of those elements in a concrete ideological discourse.' (ibid., p. 99)

The object of analysis is, then, the specificity of communication and signifying practices, not as a wholly autonomous field, but in its complex articulations with questions of class, ideology and power, where social structures are conceived as also the social foundations of language, consciousness and meaning. This is to return to prominence, along with specifically textual analysis, the questions raised by Bernstein and Bourdieu (and readdressed by Willemen) as to the structural conditions which generate different cultural and ideological competencies.

Exactly what is the nature of the 'fit' between, say, class, socio-economic or educational position and cultural/interpretative code is the key question at issue. Parkin's (1973) treatment of class structures as the basis of different meaning systems is a fruitful if crude point of departure in this respect, and provided some of the ground work for the characterisation of hypothetical/typical decoding positions as developed by Stuart Hall. I am, of course, aware of the limitations of the conceptual framework outlined by Parkin; as I have argued in 'Reconceptualising the Audience', his framework constitutes, in fact, only 'a logical rather than a sociological statement of the problem . . . providing us with the notion that a given section of the audience either shares, partly shares, or does not share the dominant code in which the message has been transmitted.'

The crucial development from this perspective has been the attempt to translate Parkin's three broad 'ideal types' of ideological frameworks (dominant, negotiated and oppositional) into a more adequately differentiated model of actual decoding positions within the media audience. Much of the important work in this respect consists in differentiating Parkin's catchall category of 'negotiated code' into a set of empirically grounded sub-variants of this basic category, which can illuminate the sociological work on different forms of sectional and corporate consciousness (cf. Beynon 1973, Parkin (op. cit.), Mann 1973, Moorehouse and Chamberlain 1974a and 1974b, Nicholls and Armstrong 1976; see also Hall and Jefferson 1978, for a 'repertoire of negotiations and responses').

Parkin's schema is, in fact, inadequate in so far as:

a) it over-simplifies the number of 'meaning-systems' in play. He locates only three: dominant, negotiated, oppositional.
b) for each 'meaning-system' he locates only one source of origin.
c) these sources of origin are derived, in each case, from different levels of the social formation.

In respect of a) it simply needs to be stated that any adequate schema will need to address itself to the multiplicity of discourses at play within the social formation; or at least, to provide systematic differentiations within the categories provided by Parkin, locating different 'versions' or inflections of each major discourse.

In respect of b) we must admit of the varied possible origins of discourses and of the contradictions within these discourses. Thus, for example, Parkin's category of a 'dominant meaning system', much in the manner of Althusser's (1971) ISAs essay, neglects the problem of contradictions within any dominant ideology. Moreover such an ideology, or rather different versions of it, may originate in a number of different places: from the media (is this what is implied by Parkin's abstract concept of the 'dominant normative order'?), from a political party, an industrial organisation, a consumer association, and so on.

In respect of c) we must put into question the 'necessity' implied by the links Parkin makes between these meaning systems and their points of origin. Thus, as argued, versions of a dominant ideology do not only originate in the media; nor do the media produce exclusively such an ideology. A 'negotiated' or subordinate meaning system can arise out of a trade union just as easily as out of a 'local working class community'; nor do such 'communities' only produce 'subordinate' ideologies, but also, on occasion (*pace* Parkin's Leninist assumptions) spontaneously radical ideologies. Thus 'oppositional' ideologies are not produced only by the 'mass political party based on the working class' but also elsewhere in the social formation. Moreover we must attend, as Parkin fails to do, to the multiplicity of ideologies produced by the range of political parties and organisations within civil society. This is precisely to focus on the relation of social and economic structure to ideology, and on the forms of articulation of the one through the other. These questions of social structure are fundamentally interlinked with the question of the differential interpretation, or decoding, of texts.

We would stress here that, whatever shortcomings Parkin's schema may have, it does allow us to conceive of a socially structured audience, and as such constitutes a considerable advance on any model which conceives of the audience as an unstructured aggregate of 'individuals' (or 'subjects'). How actual readings correspond to various types of class consciousness and social position remains to be elucidated. This is a matter for concrete investigation. What is assumed here is the viability of the attempt to relate (in however complex a form) social or class positions and decoding frameworks.

4 Problems of Technique and Method

> 'There has been a tendency in Britain for the single-minded and extremely important investigation of concepts like "film language" – and the question of the mode of production of meaning specific to the cinema ... to be accompanied by an ill-considered and unhelpful assumption that all attempts at sociological research into the conditions of reception or consumption of filmic texts can only be manifestations of the rashest empiricism. The baby (the study of the conditions of consumption) has too often been thrown out with the bathwater (the inadequacy of many of the methods employed).' (S. Harvey, *May '68 & Film Culture,* p. 78)

The above text is quoted by way of acknowledging the problems of method that remain unresolved in this work. As is evident, we are committed to a position that insists on the necessity of the 'empirical dialogue', rather than the attempt to deduce audience positions or decodings from the text in an *a priori* fashion (see Thompson (1978) p. 196, on the confusion of empirical with empiricist). However, serious problems remain in the attempt to develop an adequate methodology for the investigation of these issues – problems which have only been dealt with here in a provisional fashion. What follows is an outline of the makeshift strategies which have been adopted in the course of this project in an attempt to pursue these problems.

Bearing these points in mind, the overall plan of the research project can be seen to have been adapted from that proposed by Umberto Eco (1972).

Phases of Research

1) Theoretical clarification and definition of the concepts and methods to be used on the research.

2) Analysis of messages attempting to elucidate the basic codes of meaning to which they refer, the recurrent patterns and structures in the messages, the ideology implicit in the concepts and categories via which the messages are transmitted. (An account of the substantive products of these phases of the research can be found in *Everyday Television: 'Nationwide'*, along with a discussion of some of the problems of programme analysis. Space only allows a brief indication of the main outline of the methods of analysis employed there. The programmes were analysed principally in terms of the way they are constructed: how topics are articulated; how background and explanatory frameworks are mobilised, visually and verbally; how expert commentary is integrated and how discussions and interviews are monitored and conducted. The aim was not to provide a single, definitive reading

of the programmes, but to establish provisional readings of their main communicative and ideological structures. Points of specific concern were those communicative devices and strategies aimed at making the programmes' topics 'intelligible' and filling out their ramifications for the programmes' intended audiences.)

3) Field research by interview to establish how the messages previously analysed have in fact been received and interpreted by sections of the media audience in different structural positions, using as a framework for analysis the three basic ideal-typical possibilities:
a) where the audience interprets the message in terms of the same code employed by the transmitter – e.g. where both 'inhabit' the dominant ideology.
b) where the audience employs a 'negotiated' version of the code employed by the transmitter – e.g. receiver employs a negotiated version of the dominant ideology used by the transmitter to encode the message.
c) where the audience employs an 'oppositional' code to interpret the message and therefore interprets its meaning through a different code from that employed by the transmitter.

4) All the data on how the messages were received having been collected, these were compared with the analyses previously carried out on the messages, to see:
a) if some receptions showed levels of meaning in the messages which had completely escaped the notice of our analysis.
b) how the 'visibility' of different meanings related to respondents' socio-economic positions.
c) to what extent different sections of the audience did interpret the messages in different ways and to what extent they projected freely onto the message meanings they would want to find there. We might discover, for instance, that the community of users has such freedom in decoding the message as to make the influencing power of the media much weaker than one might have thought. Or just the opposite.

The Nationwide Audience Project: Research Procedure

The project aims were defined as being:
To construct a typology of the range of decodings made.
To analyse how and why they vary.
To demonstrate how different interpretations are generated.
To relate these variations to other cultural factors: what is the nature of the 'fit' between class, socio-economic or educational position and cultural or interpretative competencies/discourses/codes?

The first priority was to determine whether different sections of the audience shared, modified or rejected the ways in which topics had been encoded by the broadcasters. This involved the attempt to identify the 'lexico-referential systems' employed by broadcasters and respondents following Mills' (1939) proposals for an indexical analysis of vocabularies. He assumes that we can:

> 'locate a thinker among political and social co-ordinates by ascertaining what words his functioning vocabulary contains and what nuances of meaning and value they embody. In studying vocabularies we detect implicit evaluations and the collective patterns behind them, cues for social behaviour. A thinker's social and political rationale is implicit in his choice and use of words. Vocabularies socially canalise thought.' (pp. 434-5)

Thus, the kind of questions to be asked were: do audiences use the same words in the same ways as broadcasters when talking about aspects of the topic? Do respondents rank these aspects in the same order of priority as the broadcasters? Are there aspects of the topic not discussed by broadcasters which are specifically mentioned by respondents?

Moreover, beyond the level of vocabularies, the crucial questions are: to what extent does the audience identify with the image of itself presented to it via vox pop material (and via other, more implicit, definitions and assumptions about what the common sense/ordinary person's viewpoint on X is)? How far do the different presenters secure the popular identification to which they (implicitly) lay claim? Which sections of the audience accept which presenter styles as 'appropriate' points of identification for them? And, does acceptance or identification mean that the audience will then take over the meta-messages and frameworks of understanding within which the presenters encapsulate the reports? How much weight do Michael Barratt's 'summing up' comments on reports in *Nationwide* carry for the audience in terms of what code of connotation they then map the report onto? How far, for events of different degrees of 'distance' from their immediate situation and interests, do which sections of the audience align themselves with the 'we' assumed by the presenter/interviewer? To what extent do different sections of the audience identify with an interviewer and feel that they are 'lending' him/her their authority to interrogate figures in public life on their behalf?

Investigating Decodings: The Problem of Language

I have elsewhere argued that language must be conceived of as exercising a determining influence on the problems of individual thought and action. As MacIntyre puts it:

> 'The limits of what I can do intentionally are set by the limits of the descriptions available to me; and the descriptions available to me are those current in the social groups to which I belong. If the limits of action are the limits of description, then to analyse the ideas current in a society

(or subgroup of it) is also to discern the limits within which rational, intended action necessarily moves in that society (or subgroup).' (quoted in Morley, 1975)

In these terms, thinking is the selection and manipulation of 'available' symbolic material, and what is available to which groups is a question of the socially structured distribution of differential cultural options and competences.

As Mills argues:

> 'It is only by utilising the symbols common to his group that a thinker can think and communicate. Language, socially built and maintained, embodies implicit exhortations and social evaluations.'(op. cit., p. 433)

Mills goes on to quote Kenneth Burke:

> 'the names for things and operations smuggle in connotations of good and bad – a noun tends to carry with it a kind of invisible adjective, and a verb an invisible adverb.'

He continues:

> 'By acquiring the categories of a language, we acquire the structured 'ways' of a group, and along with language, the value-implications of those 'ways'. Our behaviour and perception, our logic and thought, come within the control of a system of language. Along with language, we acquire a set of social norms and values. A vocabulary is not merely a string of words; immanent within it are societal textures – institutional and political coordinates.'

Mills premises his argument about the social determination of thought on a modified version of Mead's concept of the 'generalised other', which is:

> 'the internalised audience with which the thinker converses: a focalised and abstracted organisation of attitudes of those implicated in the social field of behaviour and experience . . . which is socially limited and limiting . . . The audience conditions the talker; the other conditions the thinker.' (pp. 426-7)

However, Mills goes on to make the central qualification (and this is a point that would apply equally as a criticism of a concept of the Other derived from Lacan) that:

> 'I do not believe (as Mead does. . .) that the generalised other incorporates "the whole society", but rather that it stands for selected societal segments.' (p. 427)

This then is to propose a theory not only of the social and psychological, but also of the political, determinations of language and thought.

Problems of Hypothesis and Sample

I attempted to construct a sample of groups who might be expected to vary from 'dominant' through 'negotiated' to 'oppositional' frameworks of decoding. I aimed, with this sample, not only to identify the key points of difference, but also the points at which the interpretations of the different groups might overlap one with another – given that I did not assume that there was a direct and exclusive correspondence so that one group would inhabit only one code. Obviously, a crucial point here is that members of a group may inhabit areas of different codes which they operationalise in different situations and conversely, different groups may have access to the same codes, though perhaps in different forms.

The research project was designed to explore the hypotheses that decodings might be expected to vary with:

1) *Basic socio-demographic factors:* position in the structures of age, sex, race and class.

2) *Involvement in various forms of cultural frameworks and identifications,* either at the level of formal structures and institutions such as trade unions, political parties, or different sections of the educational system; or at an informal level in terms of involvements in different sub-cultures such as youth or student cultures or those based on racial and cultural minorities.

Evidently, given a rejection of forms of mechanistic determination, it is at this second level that the main concerns are focused. However, the investigation of the relations between levels 1) and 2), and their relations to patterns of decoding, remains important in so far as it allows one to examine, or at least outline, the extent to which these basic socio-demographic factors can be seen to structure and pattern, if not straightforwardly determine, the patterns of access to the second level of cultural and ideological frameworks.

Further, it was necessary to investigate the extent to which decodings varied with:

3) *Topic:* principally in terms of whether the topics treated are distant or 'abstract' in relation to particular groups' own experience and alternative sources of information and perspective, as opposed to those which are situated for them more concretely. Here the project aimed to develop the work of Parkin (1973), Mann (1973) and others, on 'abstract' and 'situated' levels of consciousness. The thesis of these writers is that working class consciousness is often characterised by an 'acceptance' of dominant ideological frameworks at an abstract level, combined with a tendency at a concrete, situated level to modify and re-interpret the abstractly dominant frameworks in line with localised meaning systems erected on the basis of specific social experiences. In short, this oscillation in consciousness or conception of contradictions between levels of consciousness is the grounding of the notion of a 'negotiated' code or ideology, which is subordinated, but not fully incorporated, by a dominant ideological framework.

What we need to know is precisely what kind of difference it makes to the decoding of messages when the decoder has direct experience of the events being portrayed by the media, as compared to a situation in which the media account is the audience's only contact with the event? Does direct experience, or access to an alternative account to that presented by the media, lead to a tendency towards a negotiated or oppositional decoding of the message? If so, might any such tendencies be only short lived, or apply only to the decoding of some kinds of messages – for instance messages about events directly concerning the decoders' own interests – or might there be some kind of 'spread' effect such that the tendency towards a negotiated or oppositional decoding applies to all, or to a wide range of messages? (In the project the clearest areas in which it became possible to explore these issues were in relation to the different student groups' interpretations of the 'students' item in Phase 1, and in the different trade union groups' interpretations of the 'unions' item in Phase 2).

A further level of variation which it had originally been hoped to explore, but from which time and lack of resources ultimately precluded me, was the level of contextual factors – that is, for instance, the extent to which decodings might vary with:

4) *Context:* of particular concern here were the differences which might arise from a situation in which a programme is decoded in an educational or work context, as compared with its decoding by the same respondents in the context of the family and home.

The absence of this dimension in the study is to be regretted and one can only hope that further research might be able to take it up; in particular, in the investigation of the process by which programmes are, for instance, initially decoded and discussed in the family and then re-discussed and re-interpreted in other contexts.

However, I would argue that this absence does not vitiate my results, in so far as I would hypothesise a more fundamental level of consistency of decodings across contexts. The difference between watching a programme in the home, as opposed to in a group at an educational institution, is a situational difference. But the question of which cultural and linguistic codes a person has available to them is a more fundamental question than the situational one. The situational variables will produce differences within the field of interpretations. But the limits of that field are determined at a deeper level, at the level of what language/codes people have available to them – which is not fundamentally changed by differences of situation. As Voloshinov (1973) puts it:

> 'The immediate social situation and its immediate social participants determine the "occasional" form and style of an utterance. The deeper layers of its structure are determined by more sustained and more basic social connections with which the speaker is in contact.' (p.87)

A connected but more serious absence in the research concerns the question of differential decodings, within the family context, between men and women. This is to move away from the traditional assumptions of the family as a non-antagonistic context of decoding and 'unit of consumption' of messages. Interest in this area had originally been stimulated by the results of a project investigating the decoding of media presentation of the Saltley Gate pickets of 1972 (results kindly made available to me by Charles Parker). That investigation showed a vast discrepancy between the accounts of the situation developed by miners who were at the Saltley picket and those of their wives who viewed the events at home on TV, and considerable difficulties for husband and wife in reconciling their respective understandings of the events. This material suggested the necessity of exploring the position of the 'housewife' as a viewer; in so far, for instance, as her position outside the wage labour economy, and her position in the family, predispose her to decodings in line with what I have defined (Morley, 1976) as the media's 'consumerist' presentation of industrial conflict.

Again, time and scarce resources prevented me from following up these important problems (indeed problems of particular relevance given the concentration of women within *Nationwide's* audience). We can only hope that the research currently being conducted by Dorothy Hobson of the Centre for Contemporary Cultural Studies and others on the situation of women in the media audience will allow some clarification of these issues.

Notes on Recent Audience Research

The most significant work done in this area from the point of view of this project is firstly the study by Piepe *et al.* (1978), *Mass Media and Cultural Relationships* and secondly, Blumler and Ewbank's study (1969), 'Trade Unionists, the Mass Media, and Unofficial Strikes'. In my view these projects, while focusing attention on some important issues, are critically flawed at both the theoretical and methodological levels.

Piepe's study is, from the point of view adopted here, flawed by its proclaimed concern with 'the question of "normative integration" and the manner of its attainment.' (p. ix) This is an adherence to what I have described in Chapter 1 as a 'normative' rather than an 'interpretative' model of the communicative process, which, despite the authors' later disclaimers, cannot deal effectively with the question of meanings. Indeed this is a lacuna perhaps reflected in the authors' decision not to analyse the interpretation of any specific media texts, but to concern themselves solely with their respondents' views on the media at a general level.

The study does raise important issues about the relation between patterns of differential decoding and patterns of social structure, in terms of class position, position in the housing market and community, etc. (pp. 34–35 *et seq.*). It insists on the necessity to relate psychologically based theories of media consumption – such as 'selective perception' – to social factors:

'to locate people's selective responses to media in collective contexts relating to class and community structures.' (p. 7)

Moreover the study confirms the findings here of a dimension of working class relationships to the media (and particularly to ITV) which is concerned with the questions both of localism and sectional consciousness among that class, and also with the extent to which the media are looked to for a subversive form of entertainment – a 'bit of fun' – which the BBC with its dominant paradigm of 'serious television' is seen as failing to provide.

However, the project is structured by a framework seemingly derived unproblematically from Bernstein's concepts of elaborated and restricted code, which are taken, too simply, to characterise the predominant decoding practices of the middle class and the working class respectively. Thus we are told that the:

> 'middle class use of television [is] selective and inner-directed, the working class pattern relatively more indiscriminate and other-directed.' (p. 44)

And while the middle class is characterised as having:

> 'selective perception. The ability to reject dissonant messages . . . Cognitive orientation to medium.'

the working class is characterised by:

> 'Blanket perception. Acceptance of dissonant messages . . . media orientation to reality.' (p. 46)

I would simply argue that not only are the class stereotypes invoked here too crude (cf. Rosen's critique of Bernstein) but that my own research shows plenty of examples of selective perception, rejection of dissonant messages and a distinctively cognitive orientation to the medium on the part of working class groups, with an equally complex set of responses and interpretations on the part of the middle class groups in the sample.

Blumler and Ewbank's research project is locked within the problematic of the media's 'effects'. The authors 'sought evidence of significant correlations between . . . respondents' opinions about certain trade union questions and the degree of their exposure to TV, radio and the press.' (p. 41) The questions at issue for them were those of 'the impact of the mass media in matters of concern to trade unions', (p. 33) and the degree of trade unionists' 'susceptibility or imperviousness to what they have seen, heard and read.' (p. 36) Their conclusion was that 'among all the forces that had helped in recent years to arouse trade unionists' concern about unofficial strikes, radio and TV had played a slight but significant part.' (p. 51)

The authors certainly do not see the media as providing any 'hypodermic effect'; they are aware that the audience may be resistant to persuasion and are also aware of the way that the message is mediated to the individual through his/her social context. Far from seeing the audience as an atomised mass of individuals they explicitly recognise that 'trade unionists do not face the barrage of media comment about their social situation in isolation', and that 'group ties can act as mediating variables, blocking or facilitating the impact of mass communicated messages.' (p. 34)

Nonetheless it is still a question of the degree of 'effect' on an audience of a 'message': the audience is seen to 'accept' or 'reject' given messages; selective perception and attendance only to 'sympathetic' communications are recognised as 'defence mechanisms' which the audience has at its disposal. But what is not recognised is the necessarily active role of the audience in the construction of the very meaning of the messages in question. The message is still seen here as a stimulus which may or may not produce a response at the level of attitudinal change, rather than as a meaningful sign vehicle (see Hall, 1973a), which must be 'decoded' by the audience before it can 'have an effect' or 'be put to a use'. The level at which the decoding operates is not explored.

Further, although audience members are recognised to be members of various groupings, the membership of which is seen to affect their relation to the message, the process of discussion and debate, implicit and explicit evaluation and counter-evaluation, is not explored. This is the collective process through which understandings and decodings are produced by an individual as a part of one or a number of critically significant groupings. Indeed, the authors go so far as to deny the possibility of investigating this process, declaring that 'media effects cannot be detected as they occur'. I would argue that without the analysis of this as a process, the 'recognition' of the significance of occupational family or subcultural groupings in the audience remains little more than a ritual bow to a problem.

The work in question is defined as a survey of individual opinions and attitudes, and the media are seen as having effect at the level of attitude and opinion confirmation or change. Marina de Camargo (1973) has written of this kind of research:

> 'In current political science, the sociologist who examines ideological material works with opinions, usually those given in interviews, which are responses to very precise questions, such as: what party do you vote for? why? etc.'

(in the case of Blumler and Ewbank: 'do you think newspapers/TV are pro-union, pro-employers, fair to both sides? Do you think they give too much, too little, about right, attention to strikes?') Camargo goes on:

> 'These opinion researchers have moved from the comprehensive concept of ideology to a far more limited concept of "opinion" . . . this appears as more "operational" and it can certainly be more easily specified; but it tends to take the whole framework of ideas within which the individuals express "opinions" as given and neutral, and unproblematic; all that requires pin-pointing is where individuals position themselves inside this

framework, or how their position has changed as a result of exposure to certain "stimuli".'

She makes the further, and from our perspective crucial, point that most of this kind of research 'measures merely the degree of acceptance or refusal of the ideological content of particular messages, with respect to quite specific beliefs or issues' while failing to touch on the level at which the ideology transmitted by the media may more effectively operate – through the structuring of discourses and the provision of frameworks of interpretation and meaning. It is at this latter level – of frameworks and structures – rather than at the level of the 'attitudes' or 'opinions' expressed within them that we have attempted to investigate the operation of ideology in this project.

It was this theoretical perspective that informed the methodological decision not to use fixed-choice questionnaires, as Blumler and Ewbank had done, nor to use formally structured individual interviews. Blumler and Ewbank presented their respondents with fixed-choice questions and recorded the 'substance' of their views (e.g. 'Yes I think TV pays too much attention to strikes') as the quantifiable data on which their conclusions could be based. I would argue that this separation of the content (a 'Yes') from the form in which it is expressed (the actual words used by the respondant to formulate his/her answer) is a crucial mistake: for it is not simply the 'substance' of the answer which is important, it is also the form of its expression which constitutes its meaning; not simply the number of 'yesses' or 'noes' to particular questions. To deal only with quantifiable content/substance is to substitute the logic of statistical correlations for the situated logic of actual responses. To be sure, quantitative procedures may be applicable to a later stage – once the 'logics in use' have been mapped – but quantitative analysis cannot, of itself, provide the analysis of these 'logics'.

Subsequent work by Blumler, on the actual words used by politicians in the course of the 1974 General Election, would suggest a recognition of these limitations and a revision of his earlier position. However, even in that study he concentrates on charting the frequency of appearance of particular words, rather than the preferred meanings suggested by their use in certain contexts, and the way forms and structures 'prefer' some meanings over others. Here the appropriate focus of study is surely to be placed on the scope of the available definitional frameworks, the logics-in-use, which assign particular topics to particular meaningful contexts.

'Different Languages': Project Methods

The inadequacy of a purely substantive approach, which assumes that it makes sense to add up all the 'yesses' and 'noes' given to a particular question by different respondants, is highlighted once we question the assumption that all these responses mean the same thing. As Deutscher (1977) puts it:

'Should we assume that a response of "yah", "da", "si", "oui", or "yes" all really mean the same thing in response to the same question? Or may there be different kinds of affirmative connotations in different languages?' (p. 244)

He goes on to make the point that:

'A simple English "no" tends to be interpreted by members of an Arabic culture as meaning "yes". A real "no" would need to be emphasised; the simple "no" indicates a desire for further negotiation. Likewise a non-emphasised "yes" will often be interpreted as a polite refusal.' (p. 244)

However, he argues, these are not simply points which relate to gross lingual differences; these same differences also exist between groups inhabiting different sections and versions of what we normally refer to as the 'same language'. As Mills puts it:

'writings get reinterpreted as they are diffused across audiences with different nuances of meaning . . . A symbol has a different meaning when interpreted by persons actualising different cultures or strata within a culture.' (Mills, op. cit., p. 435)

Dell Hymes makes the point that:

'The case is clear in bilingualism; we do not expect a Bengali using English as a fourth language for certain purposes of commerce to be influeced deeply in world view by its syntax . . . What is necessary is to realise that the monolingual situation is problematic as well. People do not all everywhere use language to the same degree, in the same situations, or for the same things.' (quoted in Deutscher, op. cit., p. 246)

Thus, in the first instance, I have worked with tapes of respondents' actual speech, rather than simply the substance of their responses, in an attempt to begin to deal with the level of forms of expression and of the degrees of 'fit' between respondents' vocabularies and forms of speech and those of the media (though this aspect of the research is still underdeveloped). For similar reasons I have dealt with open discussions rather than pre-sequenced interview schedules, attempting to impose an order of response as little as possible and indeed taking the premise that the order in which respondents ranked and spoke of issues would itself be a significant finding of the research.

The Focused Interview

The key methodological technique used has been the focused interview – designed, as Merton (1955) noted:

'to determine responses to particular communications . . . which have been previously analysed by the investigator.'

and crucially providing a means of focusing on:

> 'the subjective experiences of persons exposed to the pre-analysed situation in an effort to ascertain their definition of the situation.'

The initial stages of interviewing were non-directive; only in subsequent stages of an interview, having attempted to establish the 'frames of reference' and 'functioning vocabulary' with which respondents defined the situation, did I begin to introduce questions about the programme material based on earlier analysis of it. Again, following Merton, I attempted to do this in such a way that the specific questions introduced did not 'cut across the flow of the conversation' but rather engaged with, and tried to develop, points already raised by the respondents. The movement of the discussion was thus from open-ended prompting: e.g. 'What did you make of that item?' to more specifically structured questions: e.g. 'Did you think the use of that word to describe X was right?' The initial stages of the discussions enabled the respondents to elaborate, by way of discussing among themselves, their reconstruction of the programme, while the later stages enabled a more direct check on the impact of what, in the programme analysis, had been taken to be the significant points. In short, the strategy was to begin with the most 'naturalistic' responses, and to move progressively towards a more structured probing of hypotheses.

Group Interviews

The choice to work with groups rather than individuals (given that limitations of resources did not allow us the luxury of both) was made on the grounds that much individually based interview research is flawed by a focus on individuals as social atoms divorced from their social context.

This project's results confirm the findings of Piepe *et al.* (op. cit., p. 163) that while 'people's uses of newspapers, radio and television is varied, it is fairly uniform within subgroups.' While there is some disagreement and argument within the different groups over the decoding of particular items, the differences in decodings between the groups from the different categories is far greater than the level of difference and variation within the groups. This seems to confirm the validity of the original decision to use group discussions – feeling that the aim was to discover how interpretations were collectively constructed through talk and the interchange between respondents in the group situation – rather than to treat individuals as the autonomous repositories of a fixed set of individual 'opinions' isolated from their social context.

Here I would agree with Pollock (1976), 'Empirical Research into Public Opinion' (in Connerton (ed.), *Critical Sociology*) that it would be mistaken:

> 'to think of every individual as a monad, whose opinions crystallise and take on permanent existence in isolation, in a vacuum as it were. Realistic

opinion research [has] to come as close as possible in its methods of research to those conditions in which actual opinions are formed, held and modified.' (p. 229)

Where I would part company with Pollock is in his slide from the rejection of a naive empiricism towards a position in which an abstracted (Hegelian?) concept of 'public opinion' is proposed as the object of research as against the actual 'opinions' of individuals, and the method of investigation becomes correspondingly philosophical rather than empirical.

Analysing Interview Tapes

My concern has been to examine the actual speech-forms, the working vocabulary, implicit conceptual frameworks, strategies of formulation and their underlying logics, through which interpretations, or decodings, are constructed – in short, the mechanisms of 'cultural competences'. Since there is as yet no one adequate methodology for the analysis of complex, informal discourse I have employed a number of related strategies for the analysis of responses. At the first level I have attempted to establish the visible particularities in the lexical repertoires of the different groups – where particular terms and patterns of phrase mark off the discourses of the different groups one from another. Here it has been of particular interest to establish where, because of differences in overall perspective, the same terms can function in distinct ways within the discourses of the different groups.

At a second level I have been concerned to identify the patterns of argumentation and the manner of referring to evidence or of formulating viewpoints which different groups predominantly employ. Here, for instance, an attempt has been made to establish how the central topic areas identified in the programme analysis ('common-sense', 'individuality', 'the family', 'the nation', etc.) are formulated by the different groups. Particularly important here has been the attempt to establish the differential definitions of, on the one hand, 'common-sense', and on the other, 'good television' which are operated by the different groups as the points of reference from which evaluations of particular items or aspects of the programme are made. The difficulty here has been that of producing explications of such 'taken-for-granted' concepts. The attempt to directly probe such areas often meets with a resistance on the part of respondents, who presumably feel, along with Cicourel, that such attempts at precise definition of 'obvious' terms strips them of:

> 'the kind of vague or taken-for-granted terms and phrases they characteristically use as competent members of that group.' (quoted in Deutscher, op. cit.)

At a third level I have been concerned with the underlying cognitive or ideological premises which structure the argument and its logic. Here

Gerbner's work on proposition analysis (1964) has provided the main guide. As Gerbner defines it, the aim of this form of analysis is to make explicit the implicit propositions, assumptions or norms which underlie and make it logically acceptable to advance a particular opinion or point of view. In this way, declarative statements may be reconstructed in terms of the simple propositions which support or underpin them (e.g. in terms of a question in an interview, explicating the assumptions which are probably being held in order for it to make sense to ask that question). Thus, the implied premise of the following question (*Nationwide: Midlands Today*):

Q: 'But how will this research help us? What is it going to do for us?'

would be reconstructed as:

> 'Everyone knows most academic research is pointless. Can you establish your credentials as actually doing research which will have practical use-value?'

5 Responses to *Nationwide*

The project used as its 'baseline' the analyses of two *Nationwide* programmes, one broadcast in May 1976 (an extended analysis of which is to be found in *Everyday Television: 'Nationwide'*) covering a fairly representative sample of *Nationwide*'s characteristic topics, and a second programme, broadcast in March 1977 (a briefer outline of which can be found below, p. 96f.), which was a 'Nationwide Special' on the Budget and its economic consequences.

The first programme was shown to eighteen groups drawn from different levels of the educational system, with different social and cultural backgrounds, some in the Midlands region where the programme was broadcast, some in London. These were school-children, part-time and full-time students, in different levels of further and higher education.

The second programme was shown to eleven groups, some from different levels of the education system, but others from both trade union and management training centres, this time mainly in London. These were full- and part-time students in further and higher education, full- and part-time trade union officials and managers from banking institutions.

Our procedure was to gain entry to a situation where the group already had some existence as a social entity – at least for the duration of a course. We then arranged the discussions to slot into their respective courses and showed the videotape of the appropriate programme in the context of their established institutional setting.

The groups were mainly of between five and ten people. After the viewing of the tape, we tape-recorded the subsequent discussion (usually of about 30 minutes duration) and this was later transcribed to provide the basic data for the analysis.

Groups Interviewed

Phase 1: Nationwide *19/5/76*

Group No.	*Institution*	*Profile of Group*	*Size of Group*
1	Birmingham Polytechnic	Apprentice Engineers, Block Release HND Course. Mainly white, all male, aged 20-26; Working Class (WC)	10
2	,,	Apprentice Metallurgists, Day Release HNC Course. All white/male, 21-25; WC	6
3	,,	Apprentice Telephone Engineers, Block Release (City & Guilds). All white/male, 18-20; WC	13
4	,,	Apprentice Electricians, Block Release HNC. All white/male, 21-29; WC	13
5	Matthew Bolton Technical College	Apprentice Telecommunications Engineers, Block Release HNC. All white/male, 17-19; WC	12
6	,,	Apprentice Lab Technicians, Block Release ONC. 4 women, 9 men, all white, 17-19; WC	13
7	Birmingham University	Arts Degree Students. 3 men, 3 women, all white, 20-24; Middle Class (MC)	6
8	London College of Printing	Film/Photography Diploma Students. White/male, 24-26; MC	2
9*	Hackney College of Further Education	Full-time 'A' level Sociology Students. 8 men, 4 women, 6 white, 6 black, 17-29; WC	12
10	Christopher Wren School	Schoolboys (3rd year) studying for 'O' levels. 8 white, 5 black, 14-15; WC	13
11	Hackney College of Further Education	Full-time Commercial Studies Students. All women, mainly black, 17-26; WC	9
12	Christopher Wren School	Schoolboys (4th year) studying for 'O' levels. 3 white, 3 black, 15-16; WC	6
13	City College (E. London)	Full-time CSE General (Literacy) Students. 5 women, 2 men, all black, 17-18; WC	7
14	Philippa Fawcett College	Teacher Training Course Students. Mainly white, all women, 19-20; MC	12
15	,,	Teacher Training Course Students. All white, mainly women, 21-46; MC	6
16	Hackney College of Further Education	Full-time Community Studies Students. All women, all black, 17-19; WC	9
17	,,	Full-time Community Studies Students. All women, 3 white, 5 black, 17-19; WC	8
18	London College of Printing	Photography Diploma Students. 8 men, 3 women, all white, 19-26; MC	11

*Owing to a fault in the tape recording this group had later to be omitted from the analysis.

Phase 2: Nationwide *Budget Special 29/3/77*

19	Birmingham University	Arts Degree Students. All white, 2 women, 1 man, 19-21; MC	3
20	TUC Training College	Full-time TU Officials on in-service training. All white/male, 29-47; WC	6
21	Midland Bank Training College	Bank Managers on in-service training. 6 men, 1 woman, all white, 29-52; MC	7
22	TUC Training College	Full-time TU Officials on in-service training. All white/male, 24-64; WC	5
23	Polytechnic of Central London	Shop stewards on part-time Labour Studies Diploma Course. All white, 5 men, 2 women, 23-40; WC	7
24*	Hammer-smith College of Further Education	Part-time students on TU Studies course. 2 women, 1 man, all white, 24-32; MC	3
25	Hackney College of Further Education	Full-time 'A' level Sociology students. 6 women, 2 men, mainly black, 18-37; WC	8
26	London College of Printing	Print Management Trainees. All men, 3 white, one black, 22-39; MC	4
27	,,	Apprentice Printers. Both white/male, 18-19; WC	2
28	,,	Print Management Trainees. All male/black, 22-39; MC	5
29*	Midland Bank Training College	Bank Managers on in-service training. All white/male, 35-53; MC	9

*Owing to a fault in the tape recording these groups had later to be omitted from the analysis.

A BBC Audience Research Department Survey (1974) gave the composition of Nationwide's audience as follows:

	Audience	*Composition*	*Population as a whole*
	000's	%	%
Total	5,899	100.0	100.0
u-mc	321	5.4	6.0
l-mc	2,140	36.3	24.0
WC	3,438	58.3	70.0
Total			
males	2,772	46.1	
females	3,177	53.9	

Programme Description 19/5/76

Summary

Time	Item	Comments
00	Regional Menu	Use of identification pronouns;
02	National Menu	'we meet'/'the person who'.
	NEWS 'MIDLANDS TODAY'	
03	Shop steward at Coventry car plant sacked. Walsall firm cleared of charges of failure to protect their workers. Plessey management give ultimatum to workers on pay dispute.	A package of industrial news. All brief reports except for that on the carpet firm, which includes film, and some 'background' information.
(06)	Kidderminster carpet firm in danger of closure. Earth tremor in Stoke-on-Trent.	
	Mrs B. Carter goes back to meet the lions who attacked her in West Midlands Safari Park.	Questioned exclusively about her feelings. C.U. on facial expressions.
	NEWS	
	Cheltenham policeman praised by coroner for bravery. West Midlands Agricultural Show, Shrewsbury. 6 workers at Rolls-Royce Coventry win £200,000 on pools.	Photo stereotype of 'striking workers' redefined by commentators as 'individual success'.
13	Interview with Ralph Nader on consumer affairs.	'Devil's advocate' interview probing Nader's credibility.
15	WEATHER REPORT	Use of child's drawing.
	Report on a new invention from a Midlands College which will enable blind students to produce 3-D drawings. Report on a group of design students from Wolverhampton who've been building a 'Survival Kit' out of rubbish material.	Both items focus on the role of 'technological development': visual emphasis on machinery in CU. Implicit contrast made between obvious value of the invention in former item and the dubious value of the latter project.

NATIONAL *NATIONWIDE*

25 *NW* team members go on boat trip on the yacht 'Nationwide' on the Norfolk Broads.	Self-reflexive item: the *NW* team become the 'actors' in their own story.
28 A report on American servicemen and their families on a US base in Suffolk.	Extensive use of stereotypes of 'Englishness'/'Americanness' in report on 'invasion' of 'Little Old England'.
37 Interview with Patrick Meehan, released from jail with a free pardon after being originally convicted for murder.	Focusing on the subject's feelings. CU on facial expressions.
40 What to wear/eat/drink at the races. The *Nationwide* horse: Realin. Report on the financial problems of English racing. Interview with Clement Freud, a racehorse owner.	The 'Sport of Kings' brought to *NW* audience: a highly composite item involving studio mock up, outdoor film, graphics and studio interview.

Group 1

A group of mainly white, male apprentice engineers; non-unionised, with a skilled working class background, aged 20–26. Studying part-time in a Midlands polytechnic; predominantly 'don't know' or Conservative in political orientation.

The decoding of the programme material made by this group exhibits a number of contradictory tendencies. On the one hand, they at times identify and are able to articulate some of the presentational codes of the discourse which structure it in the direction of a preferred/dominant reading:

> 'People like Barratt must have some kind of controlling influence. They seem to be in command of the situation . . . he puts the last word in . . . it's like a summing-up sentence.'
> 'They're experts at looking average . . .'
> 'They go deeper than the impression they try to put over of themselves. They try to identify with the viewer.'

They recognise the operation of control in the interview situation:

> 'The interviewer was trying to guide him on the lines towards the . . . happiness of being out of jail . . . but when he goes back to saying he's going to try to get this grudge out of his system that's when the interviewer wasn't looking for that sort of aspect . . .'
> 'I think they were trying to get his feelings . . . but his opinions kept

coming through, which was what threw the interviewer, he didn't want it . . .'

And they express a view of the programme as not really 'theirs':

'They tend to pitch the tone of the programme at about the 40 year old "normal man's approach" . . .'

Their basic attitude towards the programme is one of cynicism and disbelief:

'I couldn't understand why the chap was sitting down in the pouring rain – he complained to the programme about being cold – I mean there's nothing to stop him going and sitting in his nice cosy armchair like everybody else.'
'I thought personally that interviewer was an idiot . . . he just kept repeating himself . . . silly questions . . .'

But despite their expressed cynicism and 'distance' they fundamentally share many of the cultural orientations and assumptions that underlie the discourse. They criticise the presentation – 'I think the interviewer was a bit of an idiot' – they are equivocal about the programme's 'neutrality': 'I think they stand fairly neutral . . . I think they try to appear to stand fairly neutral at times . . .' and they see that the programme embodies attitudes and orientations pre-given to the viewer:

'I don't think the interviewer likes him very much . . .'

But still, the codes and assumptions that they share lead them nonetheless to accept many of the interpretations of events and issues offered by the programme. They feel that the treatment of Nader is hostile, but:

'They're not in it for the money . . . Nader is extremely high-paid . . . *Nationwide* are doing it as a service . . . and they're willing to draw the line . . . say we must accept some change . . . but Nader, his attitude is, if you don't do it my way, you don't do it at all . . . he's powerful enough to close firms down . . .'

To this extent then, whatever their reservations about the programme's presentation of Nader, they accept the programme's definition of him as something of a 'maverick' figure – again returning to the question of monetary self-interest:

'Nader's in it for the money . . . it's a kind of racket . . . he says the consumer needs protecting, but the consumer will pay for it in the end . . . he goes to different extremes and causes more money to be spent, and the consumer pays the bill – does this community really need him?'

Interestingly, they reject the idea that *Nationwide* are to be seen as taking up a 'fourth-estate' position of 'defending' the audience/consumer against such pretensions as Nader's: they define *Nationwide*'s role as much more 'natural' and 'transparent':

'It's not so much defending us against people like Nader as showing . . . what people like him are like.'

They sense that the treatment of the students' design project on waste-materials has been trivialised in the presentation:

'They could have put a completely different slant on that . . . they could have made it seem like a scientific breakthrough.'
'Looking at it that way it can seem quite funny and it brushes any kind of importance it might have developed out of the window . . .'

But yet, despite these reservations about the treatment, their own position as part-time students on a practical/vocational training course and a strong suspicion of 'education' for its own sake, and 'arty-crafty' students, leads them to accept the perspective that *Nationwide* offers: they identify with the perspective offered by the interviewer.

Q: 'Do you think he's expressing your opinions?'
'Yeah – what the hell *were* they doing?'
Q: 'Is he asking the kind of question you would want to ask?'
'Yeah . . . because those students are obviouly living on taxpayers' and ratepayers' money for some kind of educational purpose and it's a fairly obvious question, "well, what the hell are they doing it for? Why are we throwing money into rubbish dumps?" '

Indeed, the item is remarkable to them for its uselessness:

'a bunch of students . . . living in Wales on a bunch of old rubbish – it was incredible.'

And they felt that in the interview the student representative totally failed to account for the project:

'A load of rubbish . . . and he couldn't answer . . .'
'Yeah he didn't answer . . .'
'He didn't give an answer . . . it's not an answer.'

As far as this group is concerned the students' project, in clear distinction from that of the 'blind' item, is precisely not useful or practical in any sense; and for this group practicality is an overriding consideration:

'If you ask me, they've got their priorities wrong – they spend far too much time with those students . . . I mean that's going to be *really useful*, isn't it?! I mean that thing with the blind people – they could possibly hold down a job if they've got a piece of equipment like that . . . the one is useful, and the other isn't.'

This perspective, shared between them and the programme, seems so natural to them that it hardly qualifies, in their view, as any particular perspective at all.

> 'They just said the obvious comment, didn't they . . .'
> '. . . he would have thought about the obvious questions that people are going to ask after seeing it, then ask them himself.'

The one section of the programme which their confidently common-sensical approach cannot accommodate is the Meehan interview. They validate the dramatic form of the presentation, the close-up interview:

> 'It tends to enhance the sincerity of the situation, because you can see the expression.'

and they argue that this is better than a conventional 'serious current affairs' presentation of such issues where because of the distanciation of the 'round table' discussion 'it becomes less real'.

Something has come over strongly through this highly charged interview, but it is unclear to them how to interpret it:

> 'They gave the impression that he, em, I don't know how to explain it, y'know, he was obviously cut up about it, em . . . it was the grudge that was prominent . . .'
> Q: 'What was the item about?'
> '. . . the injustice of the system' [laughter] . . . 'the fact that a guy had spent seven years of his life . . . through no fault of his own . . . but through misidentification . . . and what he was saying about those things going on in America, and also going on over here, I mean, you don't know, whether that's true or not . . .'
> 'It didn't come over strongly, because he only said it once [the reference to British Intelligence – DM] and it was only in passing, but it was certainly there, wasn't it, *something* . . .'

Group 2

A group of white, male, trainee metallurgists, aged 21–25, with a skilled upper working class background, studying on day-release for HNC qualifications in a Midlands polytechnic; predominantly 'don't know' or Conservative.

This group strongly state their preference for *ATV Today* over *Nationwide*:

> '*ATV Today* – oh yes, that's better.'

The preference is explained in terms of what they see as the more humorous approach of the ITV programme which is also seen as less 'reverent' than the *Nationwide*/BBC mode:

> 'On *ATV Today* Chris Tarrant [presenter] – he's the sort of "laughing

type", the "mucker" . . . it's a great laugh . . . and he sort of always tries to bring the person down, or give them a question they can't answer, and make them look, you know, as if they're wrong . . .'

It is a preference for a different mode of presentation and also a different type of item, one more squarely within the 'Sunday dreadful'/popular tradition of the working class press – the strange and miraculous. It is expressed with an acknowledgement that it's not 'news', but it is after all, 'a bit of a laugh':

> 'the sort of news they get on ATV – a bloody woman's being fired across a river like a cannon ball . . . every time she went straight into the river, that was great [laughter] . . . it's strange isn't it? What the bloody hell they call it news for I don't know . . .'

But the reservation is without weight: those items on *Nationwide* which the group pick out as the best are those that came closest to this mode – like the item with Mrs. Carter and the lion:

> 'It has more local interest. She's not really important but it's a nicer thing to put on, isn't it? Have everybody looking – you know, nobody wants to know Scotland's problems in the Midlands . . . it has got to be Midlands, hasn't it? Like Mrs. Brown's cat, stuff like that . . . it's the sort of thing that appeals, isn't it? Everybody in her street's going to sit and watch the programme . . .'

Here the sense of location (cf. the London working class groups on *London Today*) is the crucial thread – the point from which identification is possible. Thus the role of Tom Coyne:

> 'It's as though he's part of us . . . from Birmingham.'

and the significance of the item on the invention for the blind:

> 'Because it was a Midland invention – it shows the Midlands doing something for the blind.'

The Meehan item cannot be accommodated within these terms – all that comes over is a very negative reading:

> 'His back's up . . . he was bitter and he wants revenge . . .'
> 'underneath it all it appeared that he just wanted revenge of some sort . . .'

– a position which is simply reduced to the commonplace:

> 'After all, everybody's got some grudge against the system . . .'

Given their lack of purchase on the item, they are necessarily dependent on the presenter's framing of it:

> 'I didn't know what he [Meehan] was on about and . . . well, Barratt's a

national figure, so what he says, you know . . .'

The role of the presenters is accepted, if diffidently, as the best representative point of identification available:

> 'and they're [presenters] supposed to be sort of unbiased, take a back seat sort of view . . . in one respect it's similar to appealing to the authority, y'know, because he's such a well known person . . . the presenters have got to be the most authoritative 'cause you see most of them, they're sort of, the control . . . you mistrust the person they're interviewing, straight away, don't you? I mean, you don't know them, you're suspicious, you know, they're out for themselves, the interviewer isn't, he's only presenting the programme . . .'

Yet in some instances the presenters/interviewers are seen to 'step out of line' – which is how this group interpret the Nader interview:

> 'With the Nader interview he [interviewer] was putting the point of view of the people Nader was really fighting against. He wasn't putting the questions that the taxpayer and the, em, consumer groups would have asked him . . . he was really on the industrialist's side, you know . . . You know, he more or less asked him, do you think you're "an expert" and he pushed him on that point . . . you know, virtually called him an agitator . . . Nader was just defending himself . . .'
> '. . . he [interviewer] made a point of mentioning that Nader was being paid £2000 for the appearance, which seemed to give you the wrong impression of him being a consumer protection person to start off with – you distrust him before the item comes on!'

But the emphasis on money can also operate in another direction. Like group 1 there is an articulation of a certain cynical attitude to the whole programme – precisely in terms of money:

> 'It seems that people, well I don't know, that they . . . a lot of people seem as if they're in it for the money, more than anything else . . . a lot of those . . . even like the bloke who designed a T-square for blind people, you know, he'd patented it, so, you know. . . .'

Thus, when the programme is seen to provide a similarly hard-headed, cynical approach to the value of the students' 'rubbish' project, the group definitely endorse it. They, unlike some of the full-time student groups, regard the presentation of the students' project as 'fairly open-minded' and feel that the 'lack of sense' is a quality of the project, rather than of the presentation of it:

> 'he [the tutor] looked a bit of a nutter, didn't he . . . I was sort of . . . wondering – what the hell – we're paying! I've just paid for that, you know!'

'that's a reasonable question, you see – well, you know, what a waste of time and money!'

Similarly, the group have no hesitation in assuming as soon as the picture of the group of pools winners (six men with raised fists, in working clothes, outside a factory) is shown, that this is, of course:

'about a strike – it had a Rover sign in the back! It's usually what they're doing, isn't it? [laughter] . . . – first thing you think, whenever you hear of British Leyland is "who's on strike this time?" '

This is, simply, common sense.

In contrast with the shop stewards' (group 23) sense of themselves as a cohesive group of socially located 'decoders' this group return the argument to the level of the individual: there can be no general perspective, only a random collection of individual ones, because:

'They say they ask the questions that you'd like to know, but then again everybody's got different questions they'd like to ask. They say they try and ask things which will be general to everyone. When they say "some people would like to know . . ." it's probably, you know, it's, er, the bloke he's been for a drink with dinnertime and things like that – it's got to be all personal, hasn't it . . . he has to think, well, what would I like to know, and what would Fred like to know . . .'

Although there is, in terms of sexual difference, one categorisation of perspectives and decodings which they are aware of the programme working on:

'That blind one . . . you know, it gets the . . . [note heavily edited speech – DM], you know . . . the . . . women into it as well, you know . . . "ooh, isn't that nice", you know – "the blind can actually do things like that" . . . especially when you saw them walking out of the place, I thought that was really trying to get at people, you know, arm in arm . . . tears trickling down . . . you know, I really thought they were aiming at that . . .'

Group 3

An all white, all male group of trainee telephone engineers with an upper working class background, members of the Post Office Engineering Union, studying part-time in a Midlands polytechnic; predominantly 'don't know' politically, which is extended to a rejection of party politics.

The group, like group 2, express a marked preference for *ATV Today* over *Nationwide*, on the grounds that it is 'a bit more sort of informal', 'it's not so, sort of . . . formal' '. . . – that [Meehan item] seemed too formal, that interview with him' and, importantly for this group, on *ATV Today* 'the interviewers are younger', a point expanded on as they try to explain their disidentification with *Nationwide*:

> 'On this programme, they're all sort of middle aged people . . . there's ten years age difference . . . they're like 1950s sort of thing . . . but with those of us of our age . . . you know . . . times have changed that quickly that within ten years it can't be the same – what they think and what we think . . .'

The 'informality' of *ATV Today* is also associated for them with a sense of a greater realism – expressed in terms of 'seeing' how things 'actually are' for yourself – as distinct from what they see as *Nationwide*'s 'over-planned' and rehearsed style of presentation:

> '. . . the Americans: if that had been on *ATV Today* they'd always have one of their interviewers sort of on the camp – you'd see . . . you see nobody here . . . they didn't ask the people in Suffolk what they thought about it.'

They explain what they see as a model of acceptable/realistic presentation, as exemplified by *ATV Today:*

> 'There was a programme on *ATV Today* . . . a school . . . and the interviewer was going round, he was talking to them . . . as they were at work . . . so you sort of associated yourself with that interviewer as though you were talking to them . . .'
> 'And you actually saw the people at work rather than at home.'
> '– but here [blind item] that could have been planned.'
> '– what you see there . . . that way they were sitting . . . they could have been going round a script – whereas on *ATV Today* that thing is, like, informal, you know, just take things as they are . . .'

In contrast with groups 13 and 14, who single out the Meehan interview as the best/most interesting item in the programme, this group's view is that:

> 'it's quite boring really . . . about him being in jail for seven years.'

– a judgment informed by their very different sense of what constitutes 'good TV'. Their judgement is informed by the sense that this is the closest *Nationwide* comes to a 'serious' current affairs item.

For this group the Meehan interview makes little or no sense in itself; they are just confused:

> 'I'm not even sure if he was innocent . . . you know . . . it could be just him saying he was . . .'

and their confusion makes them dependent on the programme's framing of the item as the only way to make any sense of it:

> 'When he [Barratt] explained, at the end, you know, the full details . . . I mean . . . I could see, obviously, what had sort of happened . . . before

that . . . I didn't really feel anything about it, because . . . I didn't know enough of the details to say whether he was in the right, or wrong.'

Indeed, the category which Barratt provides, in which to 'place' the item, is reproduced as an explanation:

'It was, like, a "mistaken identity", wasn't it . . . ?'

– a 'mistaken identity' being a category only very recently in the wake of the George Davis case, established in the discourse of popular common sense.

As this group put it, when speaking of the presenters:

'You take what they say to be fact because they're generally . . . they generally say the news – like the news roundup at the end . . . the others you just . . . they give you all their personal opinions.'

Although this power is not monolithic in its operation – in the case of the Nader interview, while they feel that 'it's the bloke who interviews him who has more power because you never see him . . .' (incidentally an equation between 'invisibility' and 'power' made by several groups) – they still go on to reject the preferred, negative reading of Nader; as they put it:

'The chap's agitating . . . but it's a good thing – a lot needs changing [a paraphrase of Nader's own words] really – he's onto a good thing.'

When it comes to that section of the programme on the students' rubbish project and the invention for the blind students, they clearly pick up and endorse the contrasting evaluations signalled by the presenters – within the terms, again, of their respective practicality/use-value. The students on the rubbish project:

'They're misers . . . and they didn't really put over why they were doing it. Why do you wanna survive out of rubbish? Why do you wanna make knives and forks out of coathangers . . . especially when you can pinch 'em from the college canteen!'

The lack of rationale of the project is seen as a quality of the project itself, not of the programme's presentation of it. Most crucially though, in this group's view, the project is damned because 'they didn't seem sort of applied to anything.'

In striking contrast, the item on the invention for the blind students is seen as thoroughly rational and sensible – and again this is seen as a quality of the project, rather than of the presentation of it:

'They tell you . . . they express what it's doing for them.'
'I think you learn more from that interview . . .'
'. . . That one with the blind was about the most interesting . . .'
'– yeah, the blind.'

'The other one [students] was a waste of time: with the blind they've more or less got it for life, haven't they? Whereas with the students they've got a choice – they could go out and get a job, instead of roaming round rubbish dumps.'
(cf. group 1: 'they could hold down a job . . .')

But, while accepting and endorsing the dominant reading of the relative value of these two projects they go on to make a broader statement of qualification as to how relevant this discourse is to their lives, in general:

'Neither of them was really relevant to us though, you know, were they? You could say they were both vaguely interesting, but that's it, once it's finished, it's finished, it's finished, you know . . . you can think about it afterwards, and that . . . you know what I mean – you say "very good", and that's it.'

This perhaps is an instance of the kind of response to which Harris (1970) is referring when he speaks of the rational basis of working class alienation from the dominant forms of political and educational discourse. That is to say, in so far as the groups feel there is nothing they can do with the information which the media supply to them, it is hardly worth their while to have any opinion about it.

Group 4

A group of white, male, trainee electricians, with an upper working class background, studying part-time for HNC qualifications in a Midlands polytechnic; predominantly 'don't know' in political orientation.

The group are immediately able to 'deconstruct' *Nationwide*'s self-presentation. They see the presenters as:

'people who try to put over the view that they're, you know, sort of next door neighbours . . .'
'It's to create the impression that Tom Coyne sort of is your local mate from up the road that's in there on your behalf . . .'
'Do you feel that he is?'
'I think he's on the level myself.'
'I think it's easy to forget that he is, in the end, employed by the BBC . . .'

Although they are equivocal about the extent to which they accept *Nationwide*'s self-presentation they see it as a different issue for their parents:

'. . . he's had that effect on my father – I mean my dad calls him "Tom": "Tom's on the telly". When Tom says about the students, my dad thinks they're all layabouts, long hair, greasy and all this . . . when Tom says

something, our dad says "he's right, you know, Tom is." '
'I mean I'm forever telling myself, when I see them . . . hold on though, they're only doing it for the money . . . so I'm a bit cynical that way . . . but I know that . . . my mother is always sort of laughing about how he's [Tom Coyne] on a diet and he's not achieved this and that, you know.'

The group are thoroughly familiar with the programme's discourse and role structure:

'Bob Wellings always seems to be the one who not to believe on *Nationwide* . . . he's always the one with the funny sketches . . . the one who plays the fool . . .'
'It's producing Tom Coyne as a personality . . . and you know what Tom Coyne is going to act like . . .'

Moreover, they can also deconstruct *Nationwide*'s dominant mode of presentation:

'You're never just presented with something "bang" on the screen, you've always had something beforehand that prepares you for it, you know . . . anticipation is set up, and you're waiting for something to happen . . .'

Their overall orientation towards the programme remains distanced and cynical – as far as they are concerned:

'There was absolutely no point whatsoever in them spending half a day filming on the boat, you know. But they wanted to show us that they were definitely in the, er, East, because otherwise it could have been a studio with 'East' written on it . . . But you know what the Norfolk Broads are associated with, what they convey – so that was brought in . . .'
'. . . and they like to show that they'll have a go at anything . . .'
'. . . and they all get the expenses in!'

The cynicism at points becomes hostile: 'Barratt – he doesn't seem very well-informed . . .' 'he breaks into some sick smiles . . .' 'bit "smart" isn't he? . . .' while they recognise that, as a presenter, he does have a considerable power within the structure of the programme.

The Meehan item leaves them confused: 'He said he was innocent . . . you don't know why . . .' ' . . . it was just about "six years in solitary" . . .' '. . . you get no answer' and within this confusion they are dependant on Barratt's framing statement for clarification, about which they rather resentfully admit:

'Yeah . . . it makes a difference . . . the way that Barratt bloke winds you up and says this is another example of British justice . . . going wrong that we've shown on *Nationwide* . . .'

The Nader item they read fully within the preferred terms set up by the programme, with a particular emphasis on the chauvinistic nature of the reading. Nader is interpreted as a stereotypically 'big-shot American':

> 'He tends to be like most Americans – they have the gift of the gab . . . they tend to hold the microphone . . . he conveyed what he wanted to convey . . . obviously he was asked the right questions . . .'

They even interpret the questions (which most groups saw as evidently antagonistic towards Nader) as positively biased in his favour (cf. group 3):

> 'He wanted that question about the antagonism so that he could say, "no, I'm not" . . . I thought that question was specifically tailored to give him that space . . .'

With the students' 'rubbish' item they are well aware of the effect of the presentation:

> '[the presenter was] setting up the students . . . into a separate category of people . . . he's got all the respectable people . . . sitting at home, just got home from work, and it's . . . "look at that yob on the telly", it's obvious what he's [the presenter] on about, "he's a yob". . .'

However, despite this recognition there is a vital qualification:

> 'But that's what you pay for . . . if that's where the rates have gone . . .'

Again the project is seen as lacking any sense:

> 'I don't think any of those students actually answered why they did it . . .'
> 'they were just doing it, sort of thing . . . I still don't know why they did it . . .'

and this is seen as a quality of the project, rather than of the presentation, although there is in this group an undeveloped and solitary attempt to 'read sense' into the item:

> ' . . . perhaps it was to see if they could use their imagination and put it into practice . . . out of the classroom . . .'

The 'Americans' item is, like Nader, interpreted through a militantly chauvinistic version of the preferred reading, one which is resolved into a thoroughly xenophobic 'them/us' model:

> 'They don't change much . . .'
> 'They're all creeps . . .'
> 'They seem to be very, very false . . . so very "easy-going" . . .'
> 'sick . . . basically sick . . .'
> ' . . . about three times they said how "pleasant" they were . . . how good natured they were to people.'

They are confident about the interviewer's own stance:

> 'The interviewer was saying how "nice" the Americans were, but really, he meant, well, you know, they're not really nice . . . the interviewer was looking down on the Americans, while outwardly trying to look up at them, you know . . .'

The interpretation is premised on the construction of a sense of 'we' which is shared between the group and the interviewer, constructed by 'what we know', as distinct from these outsiders:

> Q: 'Do you think he [the presenter] thinks the Queen and the police are wonderful?'
> 'No . . .'
> 'He was being sarkie, wasn't he!'
> Q: 'How do you tell that?'
> 'Simply 'cos none of us think that way, do we?'
> ' . . . we, being a cynical race, know that . . .'
> ' . . . you were on the same wavelength as him . . .'
> 'The Englishman thinks of the Queen . . . sort of majestic . . . something like that, you know – that's the illusion he's trying to create of what we think but we know what we really think . . .'

However, while this identification is firmly secured at the level of national cultural identity, it is to some extent undercut by the group's distanciation from the image of *Nationwide* at the level of class:

> 'I think most people on *Nationwide* . . . the people we see presenting, they all seem to be snobs to me . . . I wouldn't say upper class, but getting on that way . . . the people who go to Newmarket and this sort of place . . . they are at the top, pretending they are at the bottom and trying to project the image that they think the people at the bottom ought to have . . .'

Here there is a clear distinction at the level of mode of address. The group endorse the preferred readings of particular items – Nader, students, Americans. But at the level of presentation they reject the image of the audience (and the tone 'appropriate' to that audience) which they see inscribed in the programme. In a class sense it is simply not 'for them':

> 'You wouldn't think anyone actually worked in factories – at that time of night: to them, teatime's at 5 o'clock and everyone's at home . . . a real middle class kind of attitude . . . things they cover like that boat, you know, sailing . . .'
> '. . . it's a middle class attitude, the sort of things they cover are what middle class people do.'
> '. . . the audience you can imagine are all office workers . . . commuters.'

Although the group do then go on to debate their relation to class and to middle class values in a way that shows the contradictions and instabilities within the group's perspective:

> 'But aren't the majority of working class aspiring to the middle class anyway . . .? Isn't it them they're aiming at?'
> 'I never actually go to the races . . . I don't expect many of us in here do, you know . . .'
> 'But you *want* to . . .'
> 'And they present it as if you could. They present it like how you want it. You *can* really if you . . . you can do if you want to . . .'
> 'But . . . you know . . . deep down inside you know . . . you can have a good go at it but you never actually get there . . .'

Finally, the group make an important qualification about their responses, one which, of course, without this particular emphasis, applies potentially to all the groups:

> 'There is one thing though . . . going back to that point about being too critical . . . I mean I was being critical . . . I was being asked to watch it and answer questions . . . but at home as a rule . . . when *Nationwide* is on . . . it's on all the time, I'm there just watching it, having me tea, like . . .'

Group 5

A group of white, male apprentice telecommunications engineers, with an upper working class background, mainly non-union, aged 17-18, studying part-time in a Midlands Technical College; predominantly 'don't knows' who again extend this to a rejection of party politics.

The group begin by distancing themselves from the programme:

> 'It's more trivial news, nothing very important . . . most of the programmes are made up of waffle.'

And they later remark of the 'boat trip' item:

> 'Anybody with any sense knows they've got to be messing about . . . the one who's supposed to be seasick: you wouldn't expect that to be true.'

As far as they are concerned *Nationwide* is not for them – it's for:

> 'the family sitting around watching . . . nobody in particular, just the whole family together . . .'

They resist the notion of the structuring power of the presenter's discourse, reducing the role to one of providing 'humour':

> 'They're there more as a personality, to add a bit of humour to it . . . Tom Coyne's there to be tongue in cheek . . .'

While they in fact retain verbatim Barratt's frame statement on the Meehan interview:

'He was on about "yet another case of wrong conviction".'

Yet the effect of the frame is reduced/denied:

'In an earlier programme they were going into the subject of wrongful convictions . . . and what he said afterwards just tied in with that . . . it came back to what had already gone before . . .'

After all, to ask questions about the effects of 'frame' statements is to raise questions outside the 'normal' discourse of TV discussion and the responses quickly close down the space and (re)treat the problem/issue as a technical one:

'He just sums up.' 'Tidies it up.'

The problem is acknowledged only, and purely, as a *textual* problem of 'linking', producing a 'good'/coherent programme:

'I think they're just doing a job like everyone else. They've got a job to do so they do it. I think it's only now and then when they feel strongly about something, they might just change it round a bit to try and influence people. Generally I think they've just got a job to do and they do it . . .'

This seems to me to be crucially overdetermined by the programme-makers' own technical/professional ideology, bracketing out all other considerations and working with the notions of 'good television/effective communication/interest'. These terms seem to be taken over in large part by the respondents and are both the ground on which attempts to raise other questions are denied/evaded and, further, set the terms of reference for what critical comment there is – thus criticisms are couched in terms of 'boring' ('not very skillful/silly/stupid questions/amateurish').

'The other programmes are better produced . . . *Nationwide* looks so amateurish . . . not very skillful . . . like that "blind" bit would have been done better in *Tomorrow's World*: they'd have made something of it . . .' (cf. groups 1 and 4: 'the interviewer wasn't very skillful . . .')

Again, like group 4, Nader is seen to be very much in control of his own interview:

'They let him say what he wanted . . . though he [presenter] didn't agree with him . . . I mean, he was giving long answers which usually means that he's got the upper hand . . .'

whereas:

'with the students, Tom Coyne, he was running the show, he was asking the questions, they were giving short answers. He was getting them on whatever he wanted to, whereas this Nader bloke was taking as much time as he wanted.'

This is seen quite simply as a result of Nader's personal skills:

'This Nader bloke has probably been in this situation many times before . . . if they'd had a better interviewer . . . he might not have got on so well . . . like, Robin Day would probably have shot him down . . .'

– an outcome they would have preferred, as they basically accept *Nationwide*'s negative characterisation of Nader.

With the student's 'rubbish' item *Nationwide's* 'common-sensical' dismissal of their project is endorsed:

'You could see that the people sitting at home would say, "what was the good of that? What's everyone paying for, the kids are there to learn: what have they learnt from that" . . . well, you couldn't see much point in it . . .'

This is so, notwithstanding their recognition of how the students' item has been constructed by the programme:

'On that programme you can't take it seriously . . . it seems a total waste of time! . . . Tom Coyne treats it as a kind of joke, tries to show it up as a joke, as a complete waste of time, to go to Wales to make plastic tents . . . he [Tom Coyne] was laughing at it . . .'

They can see that the students have been 'set-up' for a joke – and they agree with Coyne about them.

The sense of the obvious 'appropriateness' of *Nationwide*'s perspective is brought out oddly in relation to their presentation of the family in the 'Americans' item. When asked why Mrs. Pflieger is interviewed in the kitchen they first reply that:

'That's how a housewife is supposed to look.'

But they go on to turn the programme's construction of her upside down: her placing in the domestic context is so naturalised for them that they see the programme as simply showing where she 'is', rather than placing her somewhere. She is shown in the kitchen, they argue, 'because she was going on about a dishwasher and things' and the programme's specific role in articulating this cultural location is taken as a reflection of a natural fact.

Again, like the other apprentice groups, they discriminate between the 'rubbish' item and the item on the blind students on the basis of practicality: they see that the blind item had a specific appeal because of its location:

'I think it was probably on mainly because it had been done in Birmingham . . . it was specifically from Birmingham.'

But, most importantly:

> 'because the students are blind . . . in a bad situation . . . and in the second one where they're messing about with plastic bags . . . it's just wasting their time like, nothing . . . beneficial . . .'

The American item is read firmly within the dominant chauvinistic perspective:

> 'Terrible [laughter]. They speak with a funny language.'
> 'Plenty of money . . .'
> 'Whenever they go to a foreign country, like Japan . . . they've brought baseball there, and Coca Cola, and they just sort of invade it . . .'
> '. . . you're safe sending up Americans . . .'

and the implied chauvinism is of an obviously common-sensical nature.

Group 6

A group of mainly male, all white trainee laboratory technicians studying part-time in a Midlands Technical College; skilled working class background, mainly members of Association of Scientific, Technical and Managerial Staffs; predominantly 'don't know' or Labour.

The group are not at all enthusiastic about *Nationwide*:

> 'They state the obvious all the time . . . there was no point . . .'
> 'It would have been a lot more interesting with a blind interviewer and a blind cameraman.'
> 'I tell you what was a complete waste of time – Barratt messing about on that boat.'

They express a definite preference for ATV as more in tune with their interests: a preference in terms of its humorous, irreverent approach, as against what they see as the 'middle class', 'serious' nature of *Nationwide* as a BBC current affairs programme. The response is evidently overdetermined here by the educational context of this viewing dimension, but *Nationwide* is characterised as a 'general studies programme.'

By contrast, *ATV Today* is seen as:

> 'Not as false as that programme.'
> 'You have a good laugh . . .'
> 'Their sport is a lot better . . .'
> 'They're more human I reckon . . .' [= more like me?]
> '. . . it's a lot more realistic . . .'
> '. . . they put the odd bit in and make it interesting and humorous instead of serious and boring.'

ATV Today is praised for being 'local'; for covering the 'ridiculous but interesting':

> 'stuff like world record marrows growing . . . it's more relaxing . . . it's things that affect everyone . . . at least they tell you about his [i.e. the marrow grower] family background, how old he is, things like that, and his dad did it before him and his dad . . . what he used to do . . . what he's going to do with it when he's grown it . . .'

They at times criticise the inadequacies of *Nationwide* from a point of view which seems close to the criteria of serious current affairs:

> 'The students were only asked questions that, you know, had no real relevance; yet again, that bloke . . . [Meehan] no real relevance . . . they didn't ask the type of questions that an enquiry would ask, you know – who was the bloke who framed you? Why do you think he did it . . . and all that . . .'

Here the criticism is articulated in terms of *Nationwide* being too 'safe' and unchallenging:

> 'These interviewers, they want to keep their jobs . . . it's an early evening programme and they don't want to offend too many people . . . they can be too diplomatic in the way they approach things, especially Tom Coyne.'

But this does not mean that they would therefore endorse the 'serious/current affairs' approach. When asked to imagine the Meehan item on *Panorama* they say:

> 'They'd give us a case history of why he's in prison and you'd probably switch off.'

The Meehan interview is identified as 'serious' material; as such:

> 'That was the most boring thing in the programme.'

The group seem puzzled and frustrated by the item's combination of intensity and opaqueness – producing a very contradictory reading of it. The significance of the visuals (permanent close-up on face) is entirely denied:

> '. . . it didn't matter if you saw his face, they completely wasted the man's time, filming him, talking to him. A telephone conversation would have been just as effective.'

The point of *Nationwide*'s presentation of Meehan in this way (as 'suffering subject') is explicitly denied:

> 'I can't see at all what you got from that interview that couldn't have been summed up in five minutes of words . . .'

But at this point the precise effect of the *Nationwide* presentation, concentrating on the visual/subjective aspect, exerts its force:

> '. . . but there he was, sat on a sofa, shirt down to his navel, smoking a cigarette, probably had a can of beer in the other hand!'

This impression of Meehan is, in fact, largely constructed on the basis of the visuals supplied – extrapolated into visual stereotype (uncouth criminal?) – despite the fact that the explicit 'message' of the respondent is that the visuals were entirely redundant. The repression of the sense of this item – because the political background of the case is crudely edited out – leaves the group asking:

> 'Why? Did they say why?'
> 'Why had he been in?'
> 'I understand he claimed he was innocent – why was he innocent?'

The confusion here contrasts strongly with the reading of the 'blind' item where, because there is no similar repression of background/coherence the group can articulate the structure of the item in perfect clarity:

> 'The reporter explained how to use it, and the bloke who invented it explained why he invented it, and the kids . . . how they use it.'

The group are well aware of the role of the presenter:

> 'It's Barratt, he holds it together . . . a witty remark here and there, thrown in, this and that [i.e. a style, not a particular content?] . . . he's a well-known face . . . the news changes from day to day and you're glad to see something that doesn't . . .'
> '. . . it's so that you . . . walk into the room, you think "what's this?" – kind of, someone's fallen in the canal, and then you see Tom Coyne and say "Oh, it's *Nationwide*!" '

Moreover, they are aware of the structuring role of the presenters' 'frames' and questions – with the students' project:

> 'They were structuring the questions . . . They only wanted you to believe what the commentator, what the scriptwriter was thinking. They don't want you to make your own ideas on the subject . . .'
> 'They tried to channel you up in one direction.'

Indeed, as they can see:

> 'They humoured the students – because that's a good laugh. Tom Coyne asked him what the jumper was made of and he already knew.'

which does not imply a critical attitude to the programme's position, rather a tolerant benevolence:

> 'Yes, he's like that Tom, a bit thick.'

in which Coyne and the other presenters are redeemed because the group share their estimation of the student's project, which means that they endorse the preferred reading, despite seeing the mechanism of its construction:

> 'It's a joke . . . they didn't survive – because they took their own food . . . they just made a couple of shelters for the evening.'
> 'They did manage to make a new axe out of an old one!'
> 'It was just a chance to get on the telly, I suppose!'
> '. . . I mean, they [*Nationwide*] simply said – this is a load of kids – look at what they've done!'

Similarly with Nader: although they are perfectly aware that the programme 'had him set up as a bad guy', despite their awareness of this structure they endorse the preferred reading because, in their terms, Nader is:

> 'out to make himself seem bigger and better . . . he's too smooth . . . and he's in the money too. I mean the statement they paid him £2,000 – I'd like to do that for £2,000!!'

They accept *Nationwide*'s definition of the situation, they share the same cultural world, and what is common-sensically 'obvious' to *Nationwide* is, by and large, obvious to them too. The decoding, in line with the dominant structure embedded in the message, follows from the complementarity between their ideological problematic and that of *Nationwide*.

The Americans item is interpreted within the dominant chauvinist framework of the programme. Like the other apprentice groups the chauvinism seems to be reinforced by, or act as a displaced expression of, a form of class-based resentment at the Americans' affluence:

> 'They're used to a higher standard of living . . . so her husband didn't like it . . . and she'd got "dishpan hands"!! – like we're that "behind" in this country . . .'

At this point a subordinate thread of oppositional reading appears, taking off from the chauvinist form of opposition to the Americans towards an articulation of this opposition in a class perspective:

> 'They think we're all backward because we haven't all got three cars and two dishwashers and what have you, whereas their country is far more backward [opposition: redefinition] because they've got a higher percentage of unemployment and all the rest of it . . .'

This is extended to a criticism of the item's absences:

> 'They didn't tell you why the Americans were here though . . . you just see their social life: you don't see them loading up fighter bombers . . .'

'. . . they were trying to convey how they tried to live just how the locals live . . .'

But this is not the group's dominant perspective, which is again (as in groups 4, 12, etc.) constructed around an identification with the 'we' articulated by the programme as against the American 'them':

Q: 'What do you think the *Nationwide* people doing the programme think of Americans?'
'Probably hate them as much as we do . . .'

a perspective which takes up and endorses the programme's ironic presentation of the Americans as a 'simple race':

'They're saying the Americans are naive . . . thinking the police are wonderful and the royal family . . . we know our police aren't that good.' (cf. group 12: 'we know the other side'.)

The group share *Nationwide's* 'common sense'. When asked to define their picture of 'England now' it fits closely with that of the programme; they speak of:

'Inflation . . . they're going way ahead – leaping wages – I think *everybody* agrees that sort of thing because *everybody knows* what a bad state we're in.'

And although this perspective does not go unchallenged:

'Yeah. But it's not true is it – that's not everyone. No . . . sooner or later there's going to be a day of reckoning.'
'No . . . there isn't though . . . there's the chance for a few in America . . .
'No. There'll be a reckoning – some will feel the bite rather than others: and that will be that . . .'

the whole of this closing exchange is accompanied by a bored chorusing of:

'Ding-a-Ding . . . Ding-a-Ding . . . Ding-a-Ding . . .'

which is the group's final comment on this emerging moment of political radicalism.

Group 7

A mixed group of white students, with a middle class background, studying drama at a Midlands University, aged 19–25.

Like groups 14 and 15, this group identify *Nationwide* as a programme made for an audience which is clearly 'other':

'It's obviously directed at people with little concentration . . . it's got a kind of "easy" form . . . it's "variety", isn't it.'

They feel that the programme:

'tries to give you the impression of the presenters – that "Michael Barratt's a very nice guy" . . .'
'It's a way of treating it . . . they go in to be chatty. . . a lot of what you're watching is these same characters behaving in the same way . . .'
'It's meant to give the impression that we're all in this together. We're a great big happy family as a nation and we're all doing these things together . . . they've got this very personal approach . . . and they all chat to each other . . .'

But, they feel, *Nationwide* is 'about':

'what individuals are doing . . . something that's happened to the typical lower middle class or upper working class person . . . but in fact, if you watch it, you don't get to know any more about those individuals or what they are actually doing.'

So, assessed according to educational/informational criteria, *Nationwide* clearly fails to achieve what they see as the important goals for this kind of TV programme.

The group are, in fact, very critical of *Nationwide*'s chosen project – indeed of the programme's very claim to be 'nationwide':

'It really is very narrow. Even this excuse of going to East Anglia . . . there was nothing which dealt with East Anglian culture . . . I didn't hear an East Anglian accent . . . you've got American culture, under the guise of being in East Anglia. They go out and say "we're in East Anglia": so what? . . . that bloke could have been anywhere.'

But it is not only *Nationwide*'s claim to represent regional diversity which is rejected; the programme is also seen to represent a singular class perspective:

'There's all sorts of things in that programme . . . from a middle class point of view . . . the language used by all the interviewers . . . there was very little dialect, very little accent . . . most of the people had middle class accents . . . [they're] taking a middle class point of view, and it's basically for middle class people, that's how things fit in, like their view of the police. . .'

Again, like the teacher training college groups, the adherence to criteria derived from current affairs programming leads them to choose the Meehan interview as:

'the best part of the programme . . . on an interesting subject . . .'

in that it most closely resembles a 'proper' current affairs item in its subject matter. However, the group are, on that basis, very critical of *Nationwide*'s mode of presentation of the case:

> 'They kept fending him off . . . he was clearly quite ready to talk about it, and all they wanted to know was . . . they were emphasising the bitterness and the anger, because that's more sensational, more digestible than the "deep" . . . Yes it's "the man, the human surface" . . . "I saw it" . . . "what does the man feel like" . . . "Is that how a man looks after seven years solitary confinement" . . . if you see his face, it's as if he's talking to you individually . . .'
> ' . . . they ask him how he feels . . . but they don't ask him how he came to be convicted – none of what it's theoretically about. . .'

For them clearly its 'theoretical' status is of vital importance. However, they also negotiate their response to the item by reference to how they imagine the item might appear to that vast 'other' audience:

> 'What I was getting round to, this sort of interview, this sort of tone is perhaps what people, what the general public, like, that sort of TV, at that time during the day. That's more important than you or I who are not satisfied with that sort of depth!! . . .'
> 'Just because we don't like it . . . doesn't mean anyone out there doesn't like it, I mean they love it. Why shouldn't they?'

The Nader item is decoded in a directly oppositional fashion; they feel that the programme:

> 'set him up as a sophist to start with . . . he's coming over here for £2,000 . . . they're in fact saying, "Nader does this for money" – so it colours your impression of the fellow before you even see him.'

But despite this attempted 'closure':

> 'He came out very well actually. He answered him very well and turned it round. He kept it quite light and answered the question.'

With the 'rubbish' and blind items they feel that the programme suggests an implicit, contrasting valuation:

> 'One's saying, "this is a great advantage, made the life of the blind boys better", the other one was saying, "well you *can* do this. . ." but there were questions whether there was any advantage in it: "they're living off the state/we pay them grants to do that" is the kind of response that item is asking for.'

While they feel that this presentation aims to 'fit in with' a stereotype, 'a popular idea about students wasting time and money', their response is complex, because:

> 'There is in fact the sort of reality thing behind that – that this *is* maybe a

questionable thing to do with public money – because that's what kids do when you're eight or nine . . . making a camp . . . maybe there's a point there . . . you mustn't forget that altogether.'
'This is comical in that context – the students *do* actually go and build these sorts of useless things.'

Thus their own evaluation of the respective worth of the two projects is, in fact, in line with that of the programme, despite being resistant to the programme's manipulation of the student stereotype:

'To me the blind thing came across quite well . . . I thought it worthwhile . . . you don't get the same sort of thing about the students. You got a load of rubbish, and what you can do with a plastic bag, there's no really worthwhile thing to me about it.'
'Whereas the blind kids . . . it came over very well, it was very intelligently justified.'

They also make the point that it is the highly organised clarity of the blind item that makes for its impact:

'You know who the anonymous voice belongs to [the commentator] so you can relate to that . . . it brings you in and it takes you out . . . and the guy who designed it, he's the expert, he's telling you what it's about . . . and you've got the two kids who give you the human interest . . . telling you that it's useful. And then this voice sums up for us, "well this is what you should have gained from this item".'
'And the voice prepares you before you start, "this is what you're going to see, and this is what it's about . . ." '

The response to the 'Americans' item is contradictory; on the one hand they regard the item as 'so patronising' for presenting an image of 'Americanness' which consists of:

'having rubbish disposers . . . women who don't wash up.'
'it's so absurd. . . I mean, this woman to say that they've got two refrigerators and a dishwasher. How typical is that of American life? Essentially atypical. It may be typical of the kind of media image we get of Americans . . .'

On the other hand the programme's chauvinistic picture of the Americans does find some resonance:

'But there is this thing amongst Americans, it's the "quaintness" about England, about the police, they ask the time, they love the Queen, they come over and stand outside Buckingham Palace and so forth . . .'

They are aware of the 'absences' in the item:

'They don't say what the Americans think of the Rolls Royce factory or the environment, they say the Queen and the police.'

And they inhabit rather uneasily the 'we' which the programme constructs to include itself and the British audience, as against the 'invaders':

Q. 'What do you think *Nationwide* thinks about the Queen and policemen?'
'You don't know because if the Americans are really rather silly people, are really rather naive about Britain, which is at one level what they come across as, then you could say if the Americans say the Queen is a nice lady and British policemen are nice, then they're idiots and they're naive because they're not like that. But there's no way *Nationwide* will actually put across that point of view . . .'

But the response is also negotiated from a position which grants some reality to the image which *Nationwide* has presented:

'I can't at this stage raise many objections against that . . . I thought it was quite a good portrayal of how Americans view England and how they feel actually when they first come here . . . They probably change their ideas about the police . . .'

Nationwide's claim to 'represent' its audience is not altogether rejected. In identifying *Nationwide*'s own 'values' the group are situated uneasily in relation to them, only partly being able to endorse them:

'They can't be sceptical about everything . . . even in this programme . . . they did tend to praise some things . . . the blind drawing thing . . . things like that . . . they are serious about anything like old ladies, people who can't . . . can't drive or walk or something. That kind of thing . . . and probably someone hitting an old lady across the head, muggers, that he takes seriously. He comes across with a strong line . . .'
' . . . work doing anything with the blind is good, the students were rather silly in their waste of taxpayers' money, horse-racing would be a great loss to the British . . . "Social value for money" . . . "Practical groups in society . . ." '

But yet they give some credence to *Nationwide*'s claims:

'They've got Callaghan on next week and they're asking him our questions and they actually *are* our questions, because you have to send them in on a postcard, and they say "another man from such and such", and then answer it . . .'

Indeed, despite their criticisms, *Nationwide* remains, in part, an attractive programme:

'To me actually, you know, I think it's a very successful programme and I think it's very entertaining and I've never actually analysed why I like it but it is a good thing to watch, that's all.'

Group 8

A group of white male students at a college in London, studying full-time for a diploma in photography.

The group firmly reject the programme's presentational strategy, which they see as operating:

> 'by making it "jolly", which I find extremely creepy . . . "well, let's see what's going on in Norwich this week" . . . all very sort of matey . . . "old chap" . . . they all seem depressingly similar . . . they're all smiles . . .'
> 'I find I get a peculiar reaction against those individuals, those presenters, in that they tend to be particularly repellant, no matter what they say . . .'

It is a programme not for them, but, perhaps, for:

> 'teenagers . . . mothers putting kids to bed . . .'
> 'I don't think anyone really watches it for its news values . . . it's like *Titbits*, it really is, a magazine . . . *Weekend* sort of thing . . .'
> 'They're just like a tidied up version of the *News of the World* – lots of bits and pieces – and they don't want to risk offence.'

They are very aware of the structure of control in the programme, that it is Barratt who is:

> 'the one who puts the final – gives his opinion after a piece of film, and has, obviously, link sentences . . . so he puts *Nationwide*'s stand . . . on what we've just seen.'

But the programme's preferred readings are largely rejected by the group. In the case of the Meehan item they feel that the programme frames the item in such a way that it is presented simply as:

> 'a minor difficulty here in the judicial system . . . sort of, "I'm sorry about that, audience, but really it's nothing very important, it's just a little misidentification." '

Like some of the other groups in higher education, they regard the Meehan item, in so far as it qualifies on the criteria of serious current affairs, as potentially the most interesting thing in the whole programme – and as such to have been quite inadequately dealt with:

> '. . . I found it very insipid anyway.'
> 'The interviews are sort of prearranged; everybody's got something separate to say . . . when they did have something which could have been

interesting – like Meehan . . . they shied away from the things he was saying . . . so that at the end you get the announcer saying, "well, there's another case of a man wrongly convicted on identification" – which wasn't the case at all, according to, y'know, Meehan was trying to say something . . . very different from that.'
'That's putting . . . it into a nice careful protest . . . putting it into a situation where the audience won't be too alarmed by it . . . they won't feel the judicial system, the British Intelligence agents, are into . . . activities . . . that . . .'

The group feel that the sense of the political background to the case, which Meehan is clearly most concerned about, is simply repressed by the item:

'When Meehan's trying to question . . . making allegations . . . and the interviewer refuses to press it . . . he's almost ignoring what Meehan's saying . . . he almost said, "there you are, old chap, are you bitter?" . . . this is very serious – if he really isn't treating them seriously, we should want to know, we should want to know what it is Meehan's getting at . . .'

The group in fact interpret the problem posed for *Nationwide* by the Meehan item precisely in the terms of its 'inappropriateness' for the kind of genre/slot which *Nationwide* is:

'In a way, what Meehan thinks is completely foreign to their set-up . . . just by accident they just found that it's what they cheerfully term "an exclusive" . . . and they didn't really know what the hell to do with it.'

As they put it, all *Nationwide* could do was:

'to concentrate on the . . . "gosh, what a lucky chap you are, you've just been pardoned" . . . you know . . . "perhaps you're a little bit bitter" . . . understatement of the year!'
'I just feel the questions were blank . . . and his answers were trying to bring out something to the questions which were not permitting him to do that . . .'

The Nader item they read in an oppositional sense: they see and reject the implied evaluation of Nader supplied by the framing of the item, and go on to redefine the terms of the discussion to make sense of Nader's project:

'Well, before the Nader interview they very carefully slipped in some snide comment about how he'd received £2,000 . . . if they bothered to think, that's an individual like me, who's got to live, I mean . . . you don't sort of live by sort of . . . charity . . . you've got to . . . I dunno . . . you don't live on charity . . . he's running . . . a service . . . I mean . . . you can't just . . . sort of . . . I mean it's not some sort of holy guru . . . he's a practical . . . y'know . . . working man, who's just come across the

Atlantic . . . as part of a service really . . . he's got to be paid for it . . .'
(cf. apprentice groups' rejection of Nader, in terms of it being precisely *Nationwide,* and not him, who's providing 'a service'.)

They see the contrast between the way Nader and Meehan came over in their respective interviews as very much a product of Nader's professional skill and expertise:

'Nader himself is capable of manipulating interview time . . . he's a very fluent individual . . . and very coherent . . . presents his views . . .; with Meehan . . . the interviewer was coping with a man who for six years has been virtually in solitary confinement . . . but I don't think that exactly develops your . . . y'know . . . your coherence . . .'

And they see Nader as successfully defeating the hostile interviewer's strategy:

'Yes he does . . . I thought he did . . .'
'He more or less said . . . he said exactly what he felt agitation meant, then he answered the question . . . I mean he's really very careful . . . I mean he's . . . issued about five disclaimers and then accepted the deal . . .'

The Americans item is unhesitatingly dismissed, precisely because of its stereotyped chauvinism:

'I couldn't believe it when we got to the Americans . . . that could have come straight from 1952 or something – you know, "there are these strange people . . . called Americans . . . with strange customs; and gosh, they must find us strange too!" It was so weird to see them still put out that stuff.'
'It didn't do anything, you know, to destroy your preconceptions that have been floating around since the war . . . before that.'
'. . . and then you get some housewife talking about all her consumer luxuries . . . Christ knows how many American people are like that! It's just . . . trying to reinforce our view of . . . or misview of Americans and Americanism, you know, having those children sing "God Bless America"! This is an air-force family and they are a race apart virtually . . . I mean, I've met them in various countries throughout the world and they are very different from many other Americans . . .'

And the one aspect of the item which they feel has any real substance is, they feel, evaded:

'I mean, they come across one controversial issue . . . the Black Market, you know, which in fact is a very serious concern – the American Forces Black Market – the way it operates has the most extraordinary effects upon the communities in which it's set up . . .'

With the students' 'rubbish' item they feel that, because of the form of the questions and the presentation:

' "some students wasting the taxpayers' money" is the message that got across . . .'

But at the same time they endorse the programme's message:

'That's a weird item of no interest to anybody . . . I would have thought . . . it's a "curiosity item" . . .'

Whereas the blind students' item:

'That would find a lot of interest . . . I don't know about you, but I've sat on a bus and thought, Christ, what does he [a blind person] see inside his head . . . and really it is quite surprising . . . those drawings were extraordinarily interesting . . .'
'To me, the most interesting thing was those two blind guys saying what they got out of it . . .'

They regard their position, as students themselves, in relation to the 'rubbish' project as, indeed, raising complex questions about 'identification':

'A lot of students don't identify as such, because if you're not something different . . . not a race apart, often you would be thinking of yourself as "them" rather than "us". I mean, do you go around thinking "Christ, I'm a student!" . . . so the student audience can sit there thinking, "God, they're wasting the taxpayers' money" – forgetting the fact that he himself is a student.'

The group go further; rather than 'forgetting' their student status they, while acknowledging it, still endorse the item's valuation of this particular project:

'But still, that student item – I am a student – but it is rather a waste of time . . .'

Group 9

Owing to a fault in the tape recording this group had later to be omitted from the analysis.

Group 10

A group of mainly white schoolboys, aged 14, with a working class background, in a West London Comprehensive; predominantly 'don't know' or Labour.

The group take up and endorse many of *Nationwide*'s own criteria of what is a 'good programme'; they say approvingly of *Nationwide* that it has:

'more breadth to it . . . it has different kinds of items, lots of different reports on different stories . . . things you don't read about in the news or the paper and they tell you all about it . . .'
'They move around a bit . . . go outside as well as inside . . . makes it more

interesting . . . they've got studios all over the country . . . gives them a better scope . . .'
'Makes it different . . . more interesting for a change.'

This is in clear distinction to how they see 'serious' current affairs, which is rejected as:

'It's all politics. And things like that. For grown-ups . . . *Nationwide*'s more or less of a children's just as well as the adult's programme.'
'*Panorama*'s about, more or less, the adult's . . . Grown-up politics – things like that . . .'
'It's boring, we don't understand that . . .'

Similarly the news is not appreciated as much as *Nationwide*. About the news, this group say:

'. . . they're not interested in everyday life and things like that . . . they just wanna know about Politics. The World Cup . . . they really look *above* it. It's Politics, Politics . . .'

Whereas, they see *Nationwide* as:

'a person's programme . . . a people's programme . . . and so it's more interesting . . . like the facts of what *you* can do . . .'

From this point of view they appreciate *Nationwide*'s presentation of Meehan for its immediacy and accessibility:

'You can *see* the expressions on his face.'
'You *see* the emotional state.'

This is a valued form of experience, as against that provided by the news, in which:

'you just sort of *know* the case, what's going on.'

They comment that 'on *Nationwide* you only see the man himself' – i.e. the human subject extracted from his social/political context.

They see that this is because the interviewer only wanted him to:

'talk about "inside". . . what his day was like.'
'What he did while he was in there . . .'
'They weren't talking about the case . . . just . . . about what it was like in there . . .'
'He didn't talk about what he was put in there for.'

And they are aware that this was not what Meehan wanted to discuss, that for him 'the case was the important thing', even that, on *Panorama* perhaps:

'They'd give you more information . . . they'd let him speak up, wouldn't they . . . let him speak his mind.'

But, despite these qualifications, they feel that *Nationwide* has enabled them to grasp something of what is at issue, whereas *Panorama* or the News would have alienated them. They resolve the opacity of the item into a story of a 'mistake', which they feel they have grasped:

> 'They put him in the wrong prison.'
> 'He didn't really do it, but he was in there and he was telling about how bad it was when he wasn't really meant to be in there.'

Indeed, later in the context of discussion about the shortcomings of the police force they became more definite:

> 'He wants justice, doesn't he?'
> 'He was a "pawn" wasn't he? 'Cos he was put up to making him do it . . .'
> 'Yeah . . . but he couldn't do nothing . . .'

The Nader interview produces a clearly oppositional reading: while they see that the interviewer:

> 'tends to be questioning . . . his work . . . trying to put him off . . . trying to catch him out all the while.'

this does not prejudice their own response to Nader, who they characterise as:

> 'Serious. A serious guy.'
> 'He's trying to put across certain things . . . he looks serious.'
> 'He's trying to stop it happening to others . . .'
> 'He wanted to keep it clean . . .'
> 'They didn't want to build a power-station, nuclear waste . . .'

The items on the blind and the students' 'rubbish' project they read within the terms of the programme's dominant problematic. As far as they are concerned the 'rubbish' project was:

> 'just something that came up . . . they wanted to do something . . . so they did it . . .'
> '. . . it wasn't really all that clever.'
> 'They just want to do it – they can walk home, you know.'

And *Nationwide*'s more sympathetic presentation of the blind student's project is seen as entirely justified, precisely:

> 'Cos they're blind . . . they've not got their five senses.'
> '[But] the students, they've got their limbs, they've got all their senses, they could go out and work and they can find time to mess about with rubbish . . .'

Similarly the 'Americans' item is interpreted in line with the dominant reading. They endorse (verbatim; cf. group 7) the programme's definition of them as:

'Invaders . . .'
'They're invaders . . .'
'Yeah, like a big DaDa . . .'

And they are certainly seen as naïve in their attitude to the British police – here the group enter into the programme's construction of a 'we' of informed knowledge about these matters, though qualifying it by their reservations about *Nationwide*'s ability to express what they see as the truth of the situation. The police:

'They're just, you know, like, something we've got to learn to live with . . .'
'He couldn't really come out and say . . . if they didn't like the police . . . they couldn't really say . . . you know . . . the English police are doing wrong things, you know, and they're rubbish – they could bleep him, bleep him out . . .'
(cf. groups 2 and 6)

Group 11

A group of black (mainly West Indian/African) women, aged 18–26, working class background, studying English as part of a commercial course, full-time in a F.E. College; predominantly 'don't knows' politically.

To this group *Nationwide* is all but completely irrelevant:

'It's boring . . . *Nationwide* or anything like that's too boring.'
'Once was enough.'
'It's for older folks, not for young people.'

The programme simply does not connect with their cultural world:

'We're not interested in things like that.'
'I just didn't think while I was watching it . . .'
'I'd have liked a nice film to watch – *Love Story* . . .'

This cultural distance means that the premise of *Nationwide* – the reflection of the ordinary lives of the members of the dominant white culture, which is what gives the programme 'appeal' for large sections of that audience – is what damns it for this group:

'*Nationwide*'s a bit boring . . . it's so ordinary, not like ITV . . .'
'*Nationwide* show you things about ordinary people, what they're doing, what *Today* left out.'

Like some of the white working class groups they too express some preference for the ITV equivalent as closer to their interests:

'It's mainly on Saturdays I watch. *Sale of the Century*, *University Challenge*. *Today* is more lively. It's the way they present it. And it's shorter. You're probably looking forward to *Crossroads*.'
'[*Today*] they really do tell you what it's all about . . . and you do understand it even after seeing it for a short time.'
'They bring the population into it . . . it's not all serious . . . it's a bit of fun.'
(cf. apprentices)

In the terms of Bourdieu (1970) the group simply do not possess the appropriate cultural capital to make sense of the programme; or from another angle, the programme does not possess the appropriate cultural capital to make sense to them – although the articulation of *what* it is that constitutes the 'lack', the terms in which their distance from the dominant culture is set, are themselves contradictory:

'I don't watch it, it's not interesting . . .'
'It's the way they put things over . . .'
'They should have more politics on *Nationwide*.'
Q: 'When you say "politics" what do you mean?'
'Things what happen in this country.'
'This isn't really it, is it, *Nationwide*'s just not what you think about, is it?'
'It is going on . . . it is happening.'
'Closer to home, I mean . . .'
'Oh no, I don't like politics.'
'I don't understand it, that's why I don't like it.'

Interestingly the disjunction between their cultural world and that of *Nationwide* makes them immune to the presenter's framing statements, to the point where they cease to have any bearing beyond their simple 'textual' function:

Q: 'Do you take much notice of the "frames"?'
'Not really, they're just there to introduce things . . . they don't really make remarks after them . . .'

Their response to most of the items is one of blank indifference:

Nader

'I couldn't understand a thing he was saying . . . I don't know how much he's on the level . . .'
'I don't really know what he's done . . . anything he's done.'

Students/ 'rubbish'

'I didn't really think about it.'

'I think you'd have to be interested in things like that before you think about them.'

However, in the case of Meehan they do pick up on the suppression of the political aspects of the case, and the suggestion of police malpractice:

'They put him in jail for seven years and – he didn't do anything.'
'You don't really know what happened . . .'
'They don't want to hear about it . . .'
'They don't want the public to really know, do they.'

And in the 'Americans' item they are aware of, and reject, the chauvinism of the presentation:

'They tried to put them down – they tried to put them down and say the Americans were silly . . .'
'That couple . . . they made them seem quite silly.'

Interestingly, in comparison with the white apprentice groups, this group are ambiguously situated in relation to the British 'we' which the programme constructs in contradistinction to the Americans. To the apprentices it was clear that the programme was 'calling' the Americans naive for thinking that all British policemen are wonderful (as against the reality known by that 'we'). This group are not at all sure what the white British 'we' does know about the police, as against what they know:

Q: 'Do *Nationwide* think the police are wonderful?' 'No.'
Q: 'Is he taking the mickey out of that?' 'Yeah . . .'
'Lots of English people don't think so, do they?'
'They don't think the police are so good.'
'They think the police are good.'
'Do they? I wonder . . .'

The other aspect of the programme which they pick up on, besides the issues relating to the police, is that dimension of the programme's chauvinism constituted by its exclusive attention to Britain – which they resent; the total absence of foreign or third world news:

Q: 'Why do you like *Panorama*?'
'They go abroad, it's very interesting . . . you see things that happen abroad . . .'
'*Nationwide* is just this country . . .'
'I watch the news . . . if there's something interesting to see . . . like the news on Amin.'
'I like to hear what these nuts get up to . . .'
'*Nationwide* is OK for people who are interested in what's going on in this country only, but there are others who like to know what's going on in the

countries outside.'

'They were quite excited about going as far as East Anglia . . .'

Group 12

A group of 15 year old schoolboys, 50:50 white/West Indian, all with a working class background, in a West London Comprehensive; predominantly 'don't know' politically.

Like the apprentice groups this group frames its comments on *Nationwide* by an expressed preference for the ITV equivalent – *London Today* – as more relevant to them:

'*Today*'s better than *Nationwide* because *Today* worries about things that actually happen on the actual day they show 'em . . .'

'*Today* . . . the London programme is better for news – because *Nationwide* goes all over the place . . . but *Today* generally deals with London – that's why it's better.'

This sense of greater local identification with *Today* is also a matter of the programme's mode of address; its irreverence and 'realism' is what appeals to this group:

'*Nationwide* should have a [studio] audience . . . *Today* is better . . .'

'They keep backing off . . .'

'They should let the audience ask questions, not just the interviewer bloke . . .'

'On *Today* – take someone Royal comes on . . . they don't just talk about the things they done . . . like say they done something wrong, they talk about that, they don't try and cover them up. That's what I like about *Today* . . . it talks about the more interesting . . . like housing in London, how bad it is . . . like the punk rockers . . . the Sex Pistols, that was live – you wouldn't get that on *Nationwide*, nothing like that . . . *Nationwide* try to keep it too clean . . .'

'*Today*, you get bad stuff and things like that; *Today*'s more reality and *Nationwide*'s more like what they would like the public to hear, rather than what the public should hear . . .'

This is the hard edge of their criticism; in another sense they can also appreciate *Nationwide* as against the drier styles of 'serious television', current affairs and news:

'*Nationwide*, they bring the camera to more faces . . .'

'*Nationwide* do the more mixed sort of thing. *Nationwide* has more variety . . .'

'They're more sort of tales and myths and things like that.'

'*Nationwide*'s more for your pleasure, and the news is whatever it is . . .'

'*Nationwide*, they seem to go into more detail and historic and everything else . . .'

Thus the Meehan item is appreciated because of its immediacy/accessibility to them:

'It's better than when you're watching something like *Panorama* because you can see where it was.'
'You get all the detail.'
'You get the idea of what he went through . . . *Panorama*'s a load of rubbish, isn't it?'
(cf. group 15)
'You can see his reactions . . . see what he reacts like . . .'

Unlike many of the groups, who see the construction of the Meehan interview as obviously controlled, this group feel that far from consciously dictating the pace, the interviewer is:

'just there doing his job, isn't he?'

The interview is clear to them:

'He was just talking about . . . what happens next . . . what actions he was going to take . . . what was going to happen . . . How he was going to do it . . . what he was going to do . . .'

And they make their own sense of Meehan's statements precisely by extrapolating them to connect with their own forms of experience of the 'criminal arena', making a factually incorrect reading which retains, however, the basic point:

'Some old man got killed or some old woman, I can't remember.'
'Some break-in or something.'
'. . . he got framed up . . .'

On the other hand, they are aware that their own sympathy with Meehan may be atypical – they hypothesise at a general level a much more negotiated reading:

'After all, most people don't watch it like that – they just say, "Oh well, it's just hard luck innit?" . . .'

They are aware of the way in which the programme allocates Meehan a particular kind of 'role' – to which certain questions and not others are relevant – but they justify the 'naturalness' of the relation between the role and the questions:

Q: 'Could they interview a politician like that?'
'No, not with questions like that.'
'Because big names . . . more powerful . . . they've got to watch the questions.'
'They've got to have a bigger answer, they're not just going to stand there and say a little bit of the answer.'

Q: 'Would they ask a politician about his feelings in that way?'
'No. Obviously . . . there's no *need* to, is there?'
'. . . you're not going to do that, are you?'

Similarly the programme's presentation of Nader is clearly seen to denigrate him in a particular way – and this is justified by how 'well known' or not he is:

Q: 'Do they give him [Nader] a lot of respect?'
'No . . . they didn't, did they, because he's not known, you see . . . I doubt if many people know that lawyer, not as many as would know an MP, so they didn't treat him . . .'
'If it's someone big they couldn't, you know, give him any old questions like that . . . got to treat them, you know . . .'
'When they interview a Prime Minister or someone, they sit them in the swivel chairs, with a round table and glass of water . . . but when they interview that man they were out in the cold, and the wind was blowing, and they were standing up – they don't care about him so much . . . he isn't popular is he? If it was someone like Harold Wilson they'd treat him good wouldn't they, you know, big bloke . . .'

With the students' item they feel that the presenter's attitude to the 'rubbish' project is that:

'Well, he doesn't take it serious, the reporter doesn't take it serious . . .'
'He just shows them walking about.'
'That woman was talking about a coat, he couldn't care less about that, he just went on to something else.'
'He rushed everything, rushed through that.'
'He just went through it all and went off . . .'

And they clearly see the contrast with the presentation of the blind students:

'That was more serious.'
'Yeah . . . they did that more serious.'
'They asked more serious questions and all that.'
'It was the way they done it. They talked to the people how they felt about it . . .'

However, they go on to endorse the programme's implied valuation of the 'rubbish' project:

'They didn't say why they did it, they just said what they were made out of and all that. Didn't say the point of it all.'
Q: 'Do you think it's a serious project?'
'No, it's not important really.'
'It is if you're out in Wales on a mountain.'
'Yeah. But most people aren't though, are they?!'

The group extend their critical reading of the students' project to the point of disbelieving the factual content of the item:

'Mad. Dangerous that is . . .'
'I don't believe half of that's true – you don't find great big plastic bags in a rubbish dump . . . and where did that thread come from . . ?'

The 'Americans' item is again read squarely within the dominant/ chauvinistic framework:

'They're "friendly" . . .'
'. . . they have better homes, they're nice.'
'. . . they were proud of where they were born.'
'. . . they got luxury.'
'. . . they stuck to their own culture and they never changed their way of life.'
'. . . they're bigheads and they're lazy, but then they did it by earning it, by working for it . . .'
'. . . it was showing that what they had in America was better than what they got here. Yes, they got better luxuries all round . . . '
'They seem to be a bit . . . don't they . . . they just got big wallets, that's all! Everything's big over there . . .'

And again, the group enter into and identify with the 'we' constructed by the programme, which includes team and audience as members of a community of knowledge about the realities of British life – such as the Queen and the police force – and which precisely excludes the Americans and designates them as naïve:

'We have the lot . . . we know the other side of them [the police].'
'He [presenter] – knows the other side of it, he knows what they're like . . .'
(cf. groups 6 and 13)

This again is simply common sense, as is their attitude to industrial relations; the pools winners photo is again assumed at first to be a story about a strike:

'Cos it's outside Rolls Royce and people are always striking over there.'
'There's always strikes going on . . . there's always disputes.'
'There's always strikes going on, isn't there . . .'
'There's no point, is there – 'cos if they're going to keep striking this country's just going to go down and down . . .'

Group 13

A group of mainly female West Indian literacy students, aged 17–18, with a working class background, studying full-time in a London F.E. College.

Like group 11, this group are confused and alienated by the discourse of *Nationwide*: even the 'blind' item, which could be argued to be the clearest, most highly organised item in the programme, is seen as confusing. Although it was a relatively long item the group say that:

> 'There wasn't enough time to take it all in . . .'
> 'You don't know what they were doing . . .'
> 'It clicks . . . I don't really know . . . we know it clicks.' [i.e. the invention on the drawing board.]
> 'I still don't understand why those drawings were useful.'

However, one point does come over, perhaps the main one:

> 'I thought it was a bit amazing.'

Similarly, the Meehan item leaves the group quite confused, arguing that the missing 'key' to the item is 'a Russian bloke, some Russian bloke' who presumably is a relative of Meehan's oblique and edited reference to 'British Intelligence'. However, again, despite the confusion as to the substance of the material the 'point' – at the level of professional code values, anyway – does come over:

> Q: 'So what did come out of it?'
> 'An "exclusive interview" . . .'

The fact that this is a London group, watching a programme made in the Midlands – in the context of a course in literacy – raises the question of dialect/accent in relation to decodings, and the extent to which 'clarity' at the technical level of hearing the words interlinks with the structure of power of the discourses in the programme. They remark of Mrs. Carter (the lady interviewed with the lion):

> 'She had an accent . . . you couldn't understand her.'

The regional accents of the contributors are, then, mediated by the standard accent of the national presenter. Barratt is said to be:

> 'just making it clear to me what the guy's already said . . .'

and this is seen as an exercise of power, beyond the technical level, for:

> 'that can be dangerous actually . . . 'cos I think, "Oh, that's what it was about . . ." '
> '. . . the clearest voice presumably comes over . . . you take that as what's going on . . .'

But this sense of dependence on the presenters is mediated by a distinct lack of 'identification' with them. The presenters on the boat trip, displaying their willingness to 'have a go', are dismissed as not at all funny but rather:

'like a very poor Blue Peter item.'

Indeed, the group suggest in relation to Barratt that the programme would be:

'better off showing him eating his breakfast and shouting at his wife.'

They reject *Nationwide*'s characterisation of 'students' in the 'rubbish' item. They are well aware of the programme's construction of it, but unlike the apprentice groups, go on to reject it:

'I thought he [presenter] was awful to him [student interviewed] actually . . .'
'. . . the guy started talking and he just said "Thank you very much, that was very interesting", and he walked off . . .'
'. . . and it doesn't matter what the guy was saying but it did sound like he wasn't actually interested himself . . .'

Similarly, they reject the 'Americans' item:

'I thought the bit about the Americans was really nasty actually . . .'
'All this stuff about "nice people" . . . the way they say "nice people" made you think they were nasty people . . . it's the tone of words, isn't it?'

In particular they reject the portrayal of the family sex-roles in that item; they feel that the programme has constructed, rather than reflected, a particular image of the family and as an image of that sphere, in their experience it simply fails to fit:

Q: 'Why do they do it like that?'
'Because the wife's supposed to be in the kitchen . . .'
'But most men are in the kitchen . . . I mean they do . . .'
'It's like they're showing you a very traditional picture of how the family works: this is the woman in her place and this is the man . It's like those, what-do-you-call-it books . . . Jill and whatsisname . . .'

Group 14

This is a group of white women students, with an upper middle class background, aged 19–20, in a London Teacher Training College; predominantly 'don't know' or Conservative.

They see *Nationwide* as simply not for them. They identify it as situated in a range of domestic programming for an older, family audience:

'*Nationwide* is more for . . . general family viewing . . . like the mother rushing around getting the evening meal ready.'

The *Nationwide* audience are seen, crucially, as:

'Older . . . they're an older age . . .'
'That's the sort of programme it is – it's that time of day isn't it?'

A 'typical' *Nationwide* is seen as one where:

'they link it in with things like, I don't know, things like sort of cookery . . . and all over the country.'

It is a programme which 'I only watch with my parents.'

The programme's mode of address is alien to them, although they supply a somewhat elitist rationale for why the programme is as it is:

'I suppose at that time of day and that sort of audience, they don't want to give them anything that might force them to think or anything . . .'

The programme simply fails to 'fit' with the educational discourses in which they are situated:

'It's individually that it appeals – to people who want TV to talk to them particularly rather than a general thing like *Panorama* . . . like people on their own . . . they appeal to individuals . . . though it is a family programme as well.'

For this audience *Nationwide*:

'gives them a bit of the news – national news – and it brings it home to them . . .'
'I think the local thing is very important to whoever did the programme . . .'

The dominant criteria of evaluation are derived from that area of current affairs which does 'fit' with the discourses of their educational context. The news elements of the programme are criticised because:

'there was no detail, was there?'

In the same way the programme's presentation of Meehan is criticised by comparison with a hypothetical *Panorama* in which, they predict:

'They'd re-enact some of the case . . . and it'd be very, very detailed . . .'
'Lots of detail . . .'
'Absolutely detailed.'

– clearly a discourse in which they would feel more at home.

The personal, emotive aspects of the Meehan item, which for some groups are the best thing about it, are here seen as a poor substitute for a 'proper' treatment of the issues:

'The news is more factual . . . whereas *Nationwide*'s more personalised.'
'. . . they're more relaxed . . . they try to get over more of his personality,

showing him smoking a cig and emphasising if he touches his head or something, you know, displaying all the tension being built up . . . made you aware of his feelings . . .'

The substance of the Meehan item raises little comment in itself – except to the extent that it raises the question of *Nationwide*'s previous coverage of the George Davis case – which was seen as 'biased' in Davis' favour:

'They're biased . . . aren't they . . . when they were doing the George Davis case . . . well he's a real crook . . . and everybody got the opinion that he was so nice and everything . . . it was very biased . . .'

The presenter is seen to be in a position of power in the discourse:

'I suppose it's Michael Barratt popping up afterwards. If he grins then it's supposed to have been funny . . . if he has a straight face you're supposed to have taken it seriously.'
'They try to make their own personalities, or what they want you to see of it, show through, so that you identify with them, and more or less see what they see as the interview.'

But the strategy is less than successful; the presenters on the boat trip are found to be:

'embarrassing . . . just embarrassing . . . they went too far – it was so ridiculous.'

and Barratt himself:

'Personally, no, I don't like him . . . he can be so crushing to some people that he interviews. I find him horrible, a horrible person.'

The Nader interview is seen as misplaced in *Nationwide,* an element of 'serious broadcasting' sitting unhappily in this context:

'That was a bit too heavy for that programme.'

They clearly see *Nationwide*'s 'framing' of Nader as being quite hostile:

'They said something about him, "what would you say, because people call you the agitator?" The interviewer said that . . . You felt he was, you know . . . He couldn't defend himself very well in such a short time . . . he started off saying he got paid so much every time he talked . . . you know – he just works up a subject to get some money . . .'

But they read against the grain of the presentation and feel that Nader comes over well – because, after all:

'Well, he's a professional.'
'He's an expert. He's probably said it loads of times . . .'

Similarly with the students' 'rubbish' project, they see that the programme presents them in an unflattering way, that the interviewer:

> 'seemed to brush them off . . . I'm sure a lot of people watching would say "well, what a waste of time . . . I wonder how much it costs . . ." '

They reject this stereotypical picture of students, extrapolating what they see as the interviewer's attitude into actual dialogue:

> 'He made the remark – "this is all very jolly, but I'm looking down on you." '

And they do attempt to read 'sense' into the project; they comment that the project could have been to do with:

> 'the educational value they'd get out of this . . .'
> 'the imagination.'

although this is qualified later as a more guarded form of support for the students:

> '. . . though . . . I felt it was pretty much a waste of time . . . I mean the chap that organised it didn't come over very well . . .'

But still in the end, it is the presentation, rather than the project, which they see as at fault:

> 'There weren't enough technical questions in that at all.'
> '. . . silly questions, like why didn't you go and buy an axe, when the whole point is the destruction of society . . . that was ridiculous . . .'

Interestingly, as opposed, for instance, to the apprentice groups (who wholeheartedly take up and endorse *Nationwide*'s implied evaluation of the blind students item as much more important – because practical, useful – than the 'rubbish' project), this group reread the respective valuations. They can see that the 'rubbish' project has an 'idea' behind it and as such is of educational interest. The very practicality of the 'blind' item, its immediate, seen, usefulness is of limited interest to them:

> 'I still don't understand how that device works . . . I got the feeling . . . you don't need to understand it . . . you, just need to feel sorry for them . . .'
> '. . . it was something to do with it being useful, to them, and them talking about it. That seemed to be what they were after . . . Not an actual explanation of why or how it works.'

The preferred reading of the 'Americans' item is also rejected. While they are aware that it:

> 'tends to reinforce some British attitudes – you become very British!'

they reject the chauvinist undertones of the item. They feel that it is meant to:

'give you the impression that he thought the English were superior intellectually, even if they haven't got the equipment and the luxury . . .'

They find the presentation objectionable:

'The "nice people" . . . and it's just his whole way of speaking . . .'
'. . . he stands on the line. He's not prepared to go over it. He wants to sort of annoy people . . .'
'In a way they were laughing. When they interviewed the woman in the department store or wherever it was . . . and she was saying she respected those things . . . and I felt . . . he was laughing at the . . .'
' "Poor American Woman." '
'Yeah, and she herself was being very fair, she was being straight, and he was letting her hang herself with her own rope . . .'

Finally, they reject what they take to be the item's implied message:

'Like "what a disgraceful thing to have for breakfast". None of his business, really, what she eats for breakfast . . . Why should they blame the Americans for playing baseball in England? I mean, we go over there and play cricket! Like . . . they shouldn't be allowed to bring anything of America over with them.'

Group 15

A group of white students, predominantly women, aged between 21-46, with a middle class background, in a London Teacher Training College; predominantly Conservative.

Like group 14 this group too find *Nationwide*'s whole mode of address out of key with their own educational context: they describe the programme as:

'A muddle . . . very amateurish . . . it hopped about . . .'
'A magazine programme designed to put a smile on your face.'
'Maybe that's supposed to relieve you of any depression.'
'. . . that business on the boat . . . that was supposed to be funny . . . I thought it was really silly.'
'. . . there didn't seem to be a good reason, a valid reason, for half the things they showed . . . like that stupid thing on Americans . . .'

Here they clearly stand at the opposite end of the spectrum from those working class groups (e.g. the apprentices) who endorse *Nationwide*, but more particularly *ATV/London Today* precisely for providing variety, 'a bit of a laugh', rather than items which require 'valid reasons' for being on.

The programme's lack of coherence is problematic for them:

'The number of participants . . . from one place to another . . .'

They evaluate the programme with criteria again derived from the standards of 'serious' current affairs broadcasting; thus:

> 'There's nothing really there to . . . capture your imagination – it's not very thought provoking.'
> 'I think [it was] put out just for not the kind of people who would be interested in the "in-depth" . . . The TV equivalent of the *Sun* or *Mirror*.'
> 'It didn't feel like me . . . an older audience . . . but it's not just that.'
> 'I would have thought that Patrick Meehan and Ralph Nader were the type of people who would have been far more interesting to listen to than silly women getting mauled by lions . . .'

The programme clearly fails to achieve any sense of audience identification; as far as they are concerned the audience is clearly 'not them' – it is other people:

> 'Don't you think that those sort of people don't listen to current affairs programmes really, and if *Panorama*'s on they switch over to *Starsky and Hutch* or something . . .'

The questions the programme asks are not their questions:

> 'Perhaps they think those are the typical kind of questions that people expect . . . in which case I feel quite insulted.'

They identify and reject the programme's strategy:

> 'They're trying to bring them into your home . . . bringing the personalities into the home . . . a friendlier air, Barratt . . . and that sort of "ha, ha" . . . we're supposed to side with him . . . they're trying to get the audience more involved. Unfortunately, it does tend to have rather the adverse affect on me, because it irritates the life out of me . . . gets on your nerves after a while . . .'

From the perspective of current affairs the Meehan item is picked out as the only one meeting the appropriate criteria of substantive interest:

> 'That's the only thing it had to offer, the Patrick Meehan thing . . . really newsy . . . and interesting.'

Though they immediately go on to criticise the inadequacies of *Nationwide*'s presentation of the case:

> '. . . there could have been a lot of potential – when he said "I know they framed me, British Intelligence" that was the only bit that could have made it; if they'd gone into that . . . it was the only bit that was interesting . . . they were skirting around the subject when there's something – the bit you really want . . .'

They reject *Nationwide*'s focus on the 'human' angle:

'They asked, "what was your daily routine" – it was superficial but nothing was said about what it was really.'

They speculate that the reason for this is that:

'Perhaps his daily routine is interesting to the masses.'
Q: 'What did you get out of it?'
'Very uninformed opinions. Just an ex-con who's come out of prison . . .'
'They're trying to make it a sort of . . . human profile of a guy . . .'
'They made a big thing of it . . . exclusive . . . two hours out of prison . . . and our *Nationwide* team has been right in there . . .'

They are galled because they are aware of a missing point of coherence, something repressed in the item:

'I think Meehan really wanted to get to grips with something, didn't he? . . . keeps on saying, "But . . ." He really wanted to get onto his own thing . . . and he couldn't . . .'

The criteria of 'serious issues' taken over from current affairs is even applied by this group to the interview with Mrs. Carter about the lion. Here again the group are dissatisfied with *Nationwide*'s focus on the 'human experience' angle and sense the absence of an 'issue':

'What we wanted was somebody to ask the question *how* she came to be mauled by a lion – because at those Safari Parks you're definitely told not, on any account, to get out of the car. That wasn't taken up, was it?'

The group make a straightforward oppositional reading of the Nader interview, noting how the presentation is 'biased' against him:

'They say . . . "Nader was paid £2,000 to speak" – that's a condemnation . . . a bit of sly digging . . .'
'They asked him a funny question about what did he think of the horrible things people said about him . . .'

And this characterisation fails to fit with their own perspective on Nader:

'. . . they're undermining him, aren't they. After all, he's a very interesting man.'
Q: 'How are they undermining him?'
'By not allowing for all this good work and all the work he does put into things. They're bringing it down a bit by asking him that kind of silly question . . .'
'. . . they're always criticising him for sort of having a bash against the system, of which they are an integral part.'

The group interpret *Nationwide*'s perspective on the 'rubbish' and blind items in an unusual way:

'Both those things are really . . . optimistic . . . I think they were trying to say that our hope lies in our youth.'

But they also see an implied contrast in the evaluation of the two projects – and they themselves are more hesitant than group 14 in rejecting the implied contrast. Their own response to the 'rubbish' item is a mixture of rejection of the programme's presentation:

'the guy almost accused him of being a waste to the taxpayer.'
'like "are you wasting our money?!" '

and, at the same time, a hesitancy about the actual value of the project:

'I couldn't really see what it was they were supposed to be building . . .'
'. . . it seemed to be all play . . .'
'You couldn't really think of that as being working.'

On the other hand the response to the blind item is far less equivocal, and they endorse the programme's implied evaluation:

'Oh I think they had a different attitude to that, didn't they. They seemed to take a very serious attitude to that . . .'
'It struck me that was one of the few things on the programme that does get explained a bit. You do understand.'
'You can see the relevance of it . . . the interviewer asked the blind student . . . and he actually has a reason or a need for doing that, whereas the others don't – or it doesn't come across . . .'
'Here they get to the end and they say "Oh, here's a very worthwhile thing." '

Still, like group 14 their involvement in serious/educational discourse leads them to suggest that the item would have been better had it focused more on the 'issue' and less on the 'human' angle:

'It'd probably be best in *Tomorrow's World* – more of the machinery and less of the students – concentrate on the actual invention . . . on *Tomorrow's World* you might have a better understanding . . .'
(cf. groups 10 and 12)

In the 'Americans' item they see the programme's chauvinist orientation:

'It was very condescending . . . about their being "nice people" . . . a simple race. I think it bases the whole thing on "what are we doing with Americans in this country anyway." '

But their reading of the item negotiates uneasily between the awareness of this 'angle' and their actual response to the Americans interviewed, particularly Mrs. Pflieger:

'I think the intention was to make us feel inferior against them . . .'

'She condescends to live in a two hundred year old house. I mean, this is partly our heritage, sod it. I mean I'd like to live in places like that . . . but I know darn well I could never afford it. But she's gonna condescend to live in it . . . made me very uptight . . .'

Group 16

A group of West Indian women students, aged 18–19, with a working class background, on a community studies course full-time in a London F. E. College; predominantly 'don't know' or Labour.

To this group *Nationwide* is totally irrelevant and inaccessible:

'As soon as I see that man [Barratt] I just turn it over.'

They are so totally alienated from the discourse of the programme as a whole that they do not discriminate between the items. In the case of the blind students' and the 'rubbish' projects, which most of the white groups see as clearly contrasted within the structure of the programme, they treat both as equal:

Q: 'Were both items treated as serious stories?'
'Yeah.'
'No, it was a joke.'
Q: 'Which one?'
'Everything.'
'It's so boring, it's not interesting at all.'
'It should be banned, it's so boring.'
' . . . I think it's just like fun really watching it . . . it doesn't really interest you . . . it's like a joke or something.'
'. . . I don't see how anybody could watch it . . .'
'. . . I think I'd be asleep.'
'. . . they make it sound so serious . . . but actually to me, that was nothing . . . nothing at all . . . just crap . . .'
'I've never seen anything like that . . . I can never understand why people just sit and look at that; if I had that on my telly I'd just bang it off . . .'
'If I'd just come in . . . and that was on the telly I'd smash it up . . .'
'A really long bit of rubbish . . . rubbish . . . just telling you things you've heard already . . .'
'You watch *Nationwide* and somebody says to you – "what happened?" – and you don't know!'

The presenter's strategy of achieving an identification with the audience is less than fully successful:

'That man [Tom Coyne] sitting down around that table . . . with an umbrella . . . there was no rain and he went and opened it . . . and they show you some picture of some drops of rain [i.e. the weather report as a child's drawing] . . . I think that's really silly, that's stupid . . .'

Clearly they see *Nationwide* as a programme for a different audience, not relevant to themselves:

Q: 'Who is it for, *Nationwide*?'
'Older people – like you.'
'Middle class people . . .'
'Parents who've come in from work, especially fathers who've got nothing to eat yet.'

They are aware of the other areas of programming, however, which they see as much more relevant to themselves:

Q: 'Is there a programme meant for people like you?'
'Yes, *Today* . . .'
'*Today* isn't so bad . . .'
'*This Week* . . .'
'*World in Action* . . .'
'*Crossroads* – (Yeah).'

Their enthusiasm for this range of ITV programming is quite marked, and they see it in clear contrast to *Nationwide*:

'Those programmes bore me . . . I can't watch it – 'cos I like to watch *Crossroads*, y'know . . . '
'*Today*'s shorter . . . less boring . . . and then there's *Crossroads* on after . . .'

The style of the *Today* programme in particular is much preferred:

'*Today*'s just what happened during the day . . . one day . . . just that day, so you know exactly what has happened.'
'. . . and sometimes it has some really nice things on that you can watch . . .'

The crucial contrast between *Nationwide* and *Today* is the extent to which they fall into the category of boring, detailed, 'serious' current affairs:

'*Today* . . . they tell you what's happening and what they think about it . . .'
'*Nationwide* gets down more into detail . . . makes it more boring . . .'
'*Nationwide* . . . they go into the background – that makes it worse, 'cos it's going down further into it . . . *Nationwide* it goes right down into detail.'
'. . . *Nationwide* – they beat about the bush . . . they say it and then repeat it . . . I was so bored with it.'
(cf. teacher training groups 14, 15 for a differently accentuated use of the term 'detail' in evaluating the programme.)

The group are here not simply rejecting *Nationwide* in particular, but

Nationwide as a part of a whole range of, largely, BBC broadcasting – including the News:

' . . . when I stay and listen to the news I fall asleep.'
'. . . I can't watch it.'
'. . . I don't like the news.'
'. . . there's too much of it, you get it everywhere, in the papers, on every station on the telly . . .'

The range rejected also includes things:

'like them party political broadcasts . . . God, that's rubbish . . . those things bore me – I turn it off!'

and indeed extends to the BBC as a whole:

'I think BBC is boring.'
'BBC is really, really boring . . . they should ban one of the stations – BBC 1 or BBC 2.'

As far as this group is concerned:

'All those sort of things should be banned.'

In the absence of the power to ban the programmes, evidently the next best strategy is simply to ignore them:

Q: 'What about the interview with Nader?'
'I can't remember seeing that bit . . .'
'. . . he was there, the thing was on, so I looked . . . and I sort of listened, but it just went in and went out . . .'
Q: 'What about the Meehan interview?'
'No idea . . . not very much . . . not really . . . All I heard was that he just came out of prison . . . some murder . . . something he didn't do: that's all I heard.'

In this case, though, possibly because of a connection with their own experiences of the police, there is sufficient engagement for criticism of the item to be voiced, in some mildly oppositional sense:

'That's silliness . . . they should tell you what happened, why he was in prison, and things like that . . .'
'Anyway, he [Meehan] didn't look suspicious to me . . .'

Somewhat surprisingly, one of their most developed oppositional readings, redefining the terms of *Nationwide*'s presentation, is reserved for the piece on Mrs. Carter and the lion:

Q: 'Is everyone presented equal?'
'No: for one, that woman who went back to the tiger, I think she was an idiot.'

Q: 'Was she presented like that? Did they treat her as an idiot?'
'No, but . . .'
'. . . they did present her like . . . they took it as a joke . . .'
'But . . . he said she was a very determined person, she was very brave: "she's braver than me." '
'I don't think she was really brave, she was stupid.'

In the blind/students 'rubbish' section they read both items as equivalent, ignoring again the programme's structuration of the contrast:

'I think it was amazing . . . those people should take the chance and make all those knives and forks, you know . . .'
'All the things they made.'
'And the blind, yeah . . . really interesting . . . making something . . . they were blind but they were actually drawing like that . . .'

However, they are suspicious that these items might be there precisely to engage their interest in an underhand way:

'I think that students thing had nothing to do with education . . .'
'. . . it was interesting, though . . .'
Q: 'But do you think *Nationwide* thought it was interesting?'
'No – . . . I think they thought it was a pisstake.'
'. . . maybe they thought "oh, young people".'
'. . . it's nothing really, they sort of push that in for us – "youngsters".'

If so, they clearly see it as an inadequate 'sop' to their interest; if the programme wanted seriously to engage them:

'Why didn't they never interview Bob Marley?'

When asked what the programme would need to be about in order to interest them they reply, rather blandly:

'About everyday life . . .'
' – something that goes on.'

using the terms which *Nationwide* would indeed claim for itself: focusing precisely on their definition of the everyday. But the 'everyday life' of *Nationwide*'s mainstream culture is simply seen as irrelevant to that life as viewed by these members of a working class, black, inner city community.

Group 17

A mixed group of West Indian and white women students, aged 17–18, all with a working class background, on a full-time community studies course at a London F.E. College; predominantly Labour or Socialist.

To some extent this group respond to and validate *Nationwide*'s presentation of the domestic world:

> 'You get some good things on Nationwide . . . like they run competitions and things like that.'
> 'They do handyman of the year competition; nurse of the year . . . best neighbour of the year and all that . . . they're more interesting, those kind of things . . . like *Coronation Street* . . . at least to me.'

Moreover, they find no fault with *Nationwide*'s construction of differential 'spaces' for different statuses of contributor. They accept as commonsensical that:

> 'it seems more proper to ask a politician his views than his feelings. A politician . . . they'll ask him the practical things . . . they wouldn't ask about his feelings . . . it's what you're supposed to get, what you're trying to get from the man . . .'

They are aware of the structure of the programme, in terms of the presenter's dominance:

> 'Barratt seems to be the one most important . . . seems to do a lot of . . . the "direction changing" during the programme . . .'
> 'Always goes back to Barratt . . . . one goes through him . . .'
> 'Camera always goes back to him.'

This 'dominance' is seen only as a necessary, textual function, producing order in the discourse:

> 'I think he's got to say something short, which is roughly the centre of what it's about – something that slightly interests you – and then you're ready to watch it . . .'

This is not seen as a form of political and ideological dominance, for they say that if you happened to 'disagree' with the presenter's specific comments:

> 'if you were interested, then the things he said wouldn' worry you at all . . .'

Indeed:

> 'I don't think they do influence you.'

The presentation of Meehan is seen to focus on the subjective, emotional angle, but this is seen as a quite appropriate choice, and the suppression of the political background to the case is not taken to be a problem:

> 'I think the interviewer was interested in the feelings of the man . . . you know . . . where he'd been . . . the effects of what being inside in solitary could do to him . . . they were trying to put over how he felt, other people were shown what it would be like to be in prison . . . what prison does to people . . .'

'You could see the man . . . and tell how he's really feeling by the way he's sitting and the way his face reacts . . .'

In a similar vein *Nationwide*'s treatment of Nader is not taken to be problematic. They feel that the programme has helped him to come over sympathetically:

'He was like . . . someone doing something he feels is right . . . caring . . . . caring for people.'

clearly articulating with their understanding of the community studies course perspective.

They articulate their concerns in terms of 'youth'; this is both the sense in which they express some preference for *ATV Today*:

'I prefer *Today* . . . there's a lot more . . . it's got more things for the youth, you know.'

and the perspective from which they pick out one section of the *Nationwide* programme as particularly relevant to them:

Q: 'So is there a bit of *Nationwide* that feels like it's for you?'
'Yeah, about the students, about the lion . . .'
'The fact that they're trying to do it on their own . . .'
'Yeah . . . that was really good . . .'

Here they make their clearest break with *Nationwide*'s perspective; in so far as they see sense in the students' project:

'It was to show people that rubbish is useful– stuff can be made out of it.'

They can see clearly that the presenter is dismissive of it:

'You can see when [the interviewer] is interested in something and when he's not, 'cos when he speaks, and the type of question he asks . . . that man interviewing those people making things out of rubbish . . . you could see he wasn't interested . . . the studio had just sent him along to do it and he did it . . .'
'I know for one thing, I wouldn't have asked them [students] "what did you get from an educational point of view", I wouldn't have dreamt to ask it. I'd have said . . . you know, "did you enjoy it?" I would have said, "do you think this has got practical uses . . ." '

They see the programme's implied contrast between the 'rubbish' project and the blind students' invention as inappropriate:

'This attitude towards the rubbish is "it's nasty, dirty stuff" . . . while the blind students . . . they were really interested in that . . .'

– a sharp contrast with the apprentice groups' strong endorsement of the implied opposition between the two items.

They read the 'Americans' item within the terms of the dominant/ chauvinist problematic, seeing that the programme is being ironic about the Americans but accepting this on the grounds that they do feel themselves to belong, with *Nationwide*, to a 'we' which excludes these 'invaders':

> 'That woman [Mrs. Pfleiger] she was like a fish out of water . . .'
> Q: 'Were they taking the mickey out of the Americans?'
> 'Yeah – course they were . . . the way they phrase things . . .'
> '*Nationwide* thinks that maybe the British police and the Queen ain't that wonderful.'

– clearly an insider's knowledge that the Americans are naive not to share:

> 'I think even showing that the things we know naturally, what we accept as normal, they've got to learn, so because they do things differently, they're daft. They show them as backward – they presented them as if they were backward – dummies or something . . . he said "uncomplicated Americans", and things like that . . .'

although this identification between the group and the *Nationwide* team, as members of a community of 'natural knowledge', is to some extent qualified:

> 'I think they try and present that picture . . . but they're not really [representing us] . . .'
> 'I don't think that they could know a wide range enough of feelings, of people's feelings to be able to represent us . . .'

– a question, though, of an imperfect 'sample', or lack of comprehensiveness, not a problem of opposition of perspectives. (cf. group 2: 'it's all down to the bloke he drinks with at lunchtime . . .')

Finally, the group reject a criticism of *Nationwide* as failing to deal with serious issues like the workplace. This is seen as a form of middle-class 'serious' TV which is inappropriate. What is endorsed, as in the apprentice groups, is a notion of 'good TV' as that which gives you 'a bit of a laugh . . . variety and that.'

> 'Don't you think though, it shouldn't carry anything about work, because people do enough work during the day . . . when they come home in the evening they don't want to watch work because they do it so many hours a day . . . When they come home they want to see a bit of variety, a bit of difference . . . those extra things . . . changes from their normal life, working. If they was to show . . . Fords you'd get all the people that work for them saying, "Oh, well I've seen this – twelve hours a day already." '

Group 18

A group of mainly white, predominantly male students with a middle class background, on a full-time photography diploma course at a college in London; mainly 'don't know' politically.

The group's professional training leads them to criticise the programme from a technical point of view:

> 'I couldn't understand that bit about that guy sitting out in that tent . . . and what was the reason for having the wind in the soundtrack – that would have been easy to avoid . . .'
> '. . . I always get the feeling with *Nationwide* that somehow they never quite get it together . . . nearly every programme there's two or three mistakes . . . the next piece isn't quite ready or whatever . . .'
> 'Unless it's to make it more relaxed, watching TV make mistakes, so it's "human" . . .'

The group argue that the presenters have considerable power:

> 'They claim to speak for the viewer . . . but in doing that they're actually telling you what to think . . .'
> '. . . It's trying to be about "people", isn't it? . . . for people . . . ordinary people . . . and by saying, "a lot of people think" they're telling you what a lot of people ought to think, according to them . . . or suggesting things to people . . .'

Though the programme may not necessarily have this power in relation to them:

> 'Unfortunately, I would say that the majority of people who watch *Nationwide* accept it, accept Barratt's viewpoint, because it's easier than not . . . particularly if the audience . . . the major audience is women . . . who are probably quite hassled . . . and as they're not in any position to discuss it . . . in any other way . . . the *only* viewpoint they can take is the one in the programme – there's a chance it's the first time they've been presented with a particular argument.'

Moreover, this is an argument they apply to their own position in some cases, for instance in the discussion of the Meehan item:

> 'I don't know anything about the case, so the only interpretation I can make is that it was what Barratt says – "identification" . . .'

Because, after all:

> 'He's the voice of authority . . . he's the bloke that you see every night, sitting there telling you things.'

This power, in the case of the Meehan item, is because of the condensed and opaque form that it takes. They feel that the presentation of Meehan:

'keeps it very personal . . . just showing his face . . . his personal experience . . . just on the guy's face . . . I think they are strong visual effects . . . there's a sort of emotive series of questions . . .'
'All they want to show is how the bloke feels emotionally.'

This leaves them quite confused as to the facts of the case:

'I thought he said someone got mugged.'

and precisely because of this confusion, unwillingly dependent for whatever sense they can make of the item on the presenter's 'framing' statements.

They feel that it is the presenters who provide the cues as to the hierarchy of credibility and status of the participants; after all:

'You do it [meet the contributors] through the presenters . . .'
'Some are used for "human" stories . . . like the lions or the Americans in Suffolk and some things are "social" issues – like inventions for the blind kids.'

However, this is not to say that they necessarily accept the programme team's 'classifications' of the participants. In the case of the Nader item they feel that the programme:

'Doesn't give him much of a build-up.'

Further, that:

'. . . they make him sound a little bit eccentric.'
'I think it's significant they interviewed him outside . . . it seems less controlled and serious in approach – as opposed to if they brought him into the studios.'
'. . . they were implying that his presence [wasn't] that important – they could just go out and do a sort of "street interview".'

This is a presentation which is simply unacceptable on the basis of their prior knowledge of Nader from other sources:

'. . . and the questions too: he's a pretty important figure as far as the protection of consumers' rights is concerned. But the way they were speaking to him they were kind of cynical as to what he was going to get out of it. That was the line of questioning: "What are you doing it for?"'

And they feel that in fact, despite the hostile presentation:

'. . . he answers quite well, quite fully . . . because he could cope with that sort of situation – he's . . . been in those . . . situations very often . . .'

The group reject what they see as the chauvinism of the 'Americans' item, reading it within an oppositional framework:

'. . . the American one seems like a blatant piece of propaganda . . . for sort of . . . British Patriotism . . . laughing at the Americans . . . patronising . . . the Americans look even sillier . . . the big technology and that, they ridicule it . . . and "ho, ho, fancy having that for breakfast".'
'. . . it's ironical about the Americans . . . "they may be weird but they've got the right values . . ." '
'. . . poking fun at the Americans . . . "if they're going to live over here they ought to make the effort to be British . . . instead of playing baseball . . ." '

The group reject the programme's characterisation of the students' project, which they see as simply:

'implying that taxpayers' money is being wasted.'
'They treated it as frivolous.'
'It sort of voices the popular attitude towards students . . . you know, they said, "you all had a very exciting time" and there was a shot of this guy lying in a sleeping bag . . .'
'The interviewer also seemed to cut everybody off . . . each student, he seemed to cut them off.'

This last point – the way in which the discussion is cut about and therefore incoherent – they take to be, in fact, the most crucial one:

'The rubbish thing . . . I couldn't work out what it was meant to be.'

and they contrast that very clearly with the coherence of the 'blind' item:

'In the interview with the blind, it worked to a complete conclusion . . . you saw them walking out of the building together . . . it was more structured and planned out . . . more together . . . taking more pains to make it into a finished piece . . .'

The point here seems to be that coherence, as a prerequisite of comprehension or memorability, is only produced within the programme for some discourses, or some items, and not for others. This is to suggest a 'politics of comprehension' where 'balance' allows that other, subordinate perspectives and discourses, beyond that of the preferred reading, will 'appear' in the programme – but not in coherent form – and the dominance of the preferred reading is established, at least in part, through its greater coherence.

Phase 2: *Nationwide* 'Budget Special' 29/3/77

Phase 2 of the project, using a *Nationwide* programme on the March 1977 budget, was designed to focus more clearly on the decoding of political and economic issues, as opposed to the coverage of 'individuals' and 'social oddities' represented in the programme used in Phase 1. In particular this sample of groups was chosen so as to highlight the effects of involvement in the discourse and practice of trade unionism on decoding patterns. The groups chosen were managers, university and F.E. students, full time TU officials and one group of shop stewards.

The programme was introduced by Frank Bough, as follows:

> 'And at 6.20, what this "some now, some later" Budget will mean to you. Halma Hudson and I will be looking at how three typical families across the country will be affected. We'll be asking . . . union leader Hugh Scanlon and industrialist Ian Fraser about what the budget will mean for the economy.'

Three main sections from the programme were selected for showing to the various groups:

1) A set of vox pop interviews with afternoon shoppers in Birmingham city centre on the question of the tax system and whether:
a) taxes are too high, and
b) the tax system is too complicated.
These interviews are then followed by an extensive interview with Mr. Eric Worthington, who is introduced as 'a taxation expert'. Mr. Worthington moves from technical discussion of taxation to expound a philosophy of individualism and free enterprise and the need for tax cuts to increase 'incentives', combined with the need for cuts in public expenditure. It is notable here that the interviewer hardly interrupts the speaker at all; the interview functions as a long monologue in which the speaker is prompted rather than questioned.

2) The main section of the programme, in which *Nationwide* enquires into:

> 'how this budget will affect three typical families . . . and generally speaking most people in Britain fall into one of the three broad categories represented by our families here . . . the fortunate 10% of managers and professionals who earn over £7,000 p.a., the less fortunate bottom fifth of the population who are the low paid, earning less than £2,250 p.a., and the vast majority somewhere in the middle, earning around £3,500 p.a.'

The three families are then dealt with one at a time. Each 'case study' begins with a film report that includes a profile of the family and their economic situation, and an interview which concludes with the husbands being asked what they would like to see the Chancellor do in his Budget. Following the film report, the account then passes back to the studio where Bough and Hudson work out by how much each family is 'better off' as a result of the Budget. Each family (the husband and the wife) is then asked for its comments.

The families chosen are those of an agricultural labourer, Ken Ball, a skilled toolroom fitter, Ken Dallason, and a personnel manager, John

Tufnall. The general theme of the programme is that the budget has simply 'failed to do much' for anyone, though the plight of the personnel manager (as representative of the category of middle management) is dealt with most sympathetically.

3) The third section is again introduced by Bough:

> 'Well now, with one billion pounds' worth of Mr. Healey's tax cuts depending upon a further round of pay agreement, we are all now, whether we are members of trade unions or not, actually in the hands of the trade unions.'

There follows a discussion between Hugh Scanlon (Associated Union of Engineering Workers) and Ian Fraser (Rolls Royce), chaired by Frank Bough, which concentrates on the question of the power of the unions to dictate pay policy to the government. Here Scanlon is put on the spot by direct questions from both Ian Fraser and Frank Bough in combination, whereas Fraser is asked 'open' questions which allow him the space to define how he sees 'the responsibility of business'.

Group 19

A mixed group of white, upper middle class drama students at a Midlands University, aged 19–21; no predominant political orientation.

This group judge *Nationwide* by the criteria of 'serious' current affairs, and find the programme's whole mode of presentation inappropriate to their interests, and out of key with the educational discourses within which they are situated:

> 'It's like Blue Peter . . . it's vaguely entertaining . . . basically undemanding . . . doesn't require much concentration . . . stories that have absolutely no implications apart from the actual story itself . . .'
> ' . . . sort of dramatic . . . teatime stuff, isn't it? . . .'
> ' . . . it's all very obvious . . . here's a big board with a big chart – it's like being in a classroom: very simplified.'
> ' . . . just novelty . . . they almost send them up by their mock seriousness . . . they'd make a story out of nothing . . . someone who collects something very boring, making models out of matchsticks or something . . .'

In contradiction, however, they do later argue (contrary, for instance, to group 21, who reject this personalised mode) that the *Nationwide* form of presentation does have the advantage that:

> 'You remember a personality easier than you can a table of figures . . . I can remember those three people . . . whereas, if they'd talked about "three average families", you'd just forget all the particulars.'

The 'three families' section of the programme is seen to be aimed at getting over:

> 'what it means to these three families – trying to put a "human interest" on it . . .'
> '. . . it tends to neaten it up, makes it very accessible in a way . . . Britain is divided into three categories and you slot yourself into one of the three . . .'

However, it is a form of identification which they are not happy about making:

> 'Well I suppose you should be able to slot yourself into one of those categories, just by the amount of money your parents earn . . . em . . . whether or not you'd say they're social categories as well . . .'

However, unlike the TU groups, they do not feel that the item proposed a particular class perspective; rather:

> 'Really, it's three ways of saying that none of them are satisfied . . . just three examples that it wasn't a very good budget to anyone.'

Indeed they remark that:

> 'The whole thing was biased against the budget . . . they didn't have a single government spokesman sticking up for it at all, which I thought was a bit naughty.'

But, crucially, as far as they are concerned:

> 'Everybody's bound to have a different say.'

It is a question, for them, of individual rather than class perspectives. Thus, the vox pop interviews, which the TU groups (e.g. group 22) comment on as being exclusively middle class, are for them:

> 'seven different individuals.'

although 'class' re-emerges here in a displaced way, as indexed by 'coherence' or 'thickness':

> 'I was amazed how coherent those people were, you didn't have anyone saying, "Yer what?" You often get that.'
> '. . . if they'd have picked somebody who was terribly incoherent, a bit thick and that, we'd have criticised it for that . . .'

They reject any idea that the programme's complementarity or perspective with a particular class group leads them to communicate differently with the different families. Rather, at a general level:

> 'I felt they were all a bit patronising.'
> 'I wouldn't have said he communicated in a different way to the three families, actually.'

As they put it, class is simply a 'fact of life':

'Surely he's already accepting, maybe wrongly, that there are, obviously, differences in income . . . different classes, different educations, and he's not trying to say . . . you're all equal, he's presupposing . . . it's a fact of life. You've got three different people there, and £1 to that one is . . . all right, it's callous, but it's going to make more difference than £1 to another set. So yes, OK, in a way he's . . . it's being discriminating . . . when you analyse it down . . . but it's a fact of life . . .'

They do see some partiality in *Nationwide*'s presentation, to the extent that the programme is seen to focus on the themes of:

'Incentive to work . . .'
'Middle management . . . they kept on talking about middle management.'

and to present the respective cases in such a way that:

'They do it as if he's [i.e. the personnel manager] hard done by, because he's got his mortgage to pay, but he's also living in a nicer area . . .'

and moreover to display an important lacuna in that:

'They try to give a wide spectrum by showing those three families, but in fact they left out the fourth lot – which is anything over £7,000 . . .'

However, the strongest criticism of the programme's presentation of this item comes not from a class but from a feminist perspective, especially important as they feel that the programme is designed principally for an audience of:

'Women . . . they're the only people home at 6 o'clock.'
'The budget approach . . . all those bits about budgeting, how much housekeeping . . .'
'It's surely all directed towards women . . .'
'. . . this is very heavily guided towards just how much money the women is going to get. In all of those cases, it was always Mrs. X – there was the wife not affording this and not affording that.'
'Even the woman who goes to work . . . they say, how do you spend his money . . . but she's earning . . .'
'And they ask that woman "what's your husband's line of business" – and she breaks in, "and I am also . . ." '

Despite these criticisms, and the fact that they feel that the discourse of the programme is:

'. . . all middle . . . sort of very generalised thing . . . especially the national part . . . whereas *Midlands Today* seems a bit more . . . more . . .

well . . . because there's masses of factory workers watching it . . .'
'. . . on a general level it goes back to a BBC audience . . . general, bland . . . middle class presentation . . .'

they are unwilling to reject the overall discourse, which they feel retains its validity, for instance, in the case of the linking thread provided by the 'expert' who assesses the three families' tax/budget situations; because:

'You can't really distrust those figures, unless the guy just makes up things . . . he did have a reason for each figure, and it wasn't, em, an *opinion*, it was a *fact*.'

In the case of the tax expert interviewed at length at the beginning of the programme (Mr. Worthington), although they see him at a formal level as 'out of control' they do not see him as presenting a particular political/economic perspective:

'Well, he was just let loose . . . at least . . . he had no-one against him . . . all the time, you thought, God, this is amazing . . . he was allowing him to run riot . . . he was saying absolutely anything . . . I mean he could have blasted out "Buy Ryvitas" and all trade names . . . the chap was out of control . . . just giving a very personal point of view, saying *I, I* think . . .'

In commenting on the trade union/employer discussion they justify Bough's attack on Scanlon as reasonable 'in the situation':

'Considering the weight of the responsibility *was* with the trade union – I mean it was dependent on the trade union vote so he can't be sort of completely equal because one question is really more important in the discussion than the other question.'

This, I would argue (see Morley, 1976), is to accept the media's construction of the situation as one in which the trades unions bear principal responsibility for inflation, etc., and to fail to see that this *is* a constructed view. Indeed, they deny the process of its construction:

'I don't think they have done anything to bias us one way or another there, there's nothing they can do . . .'
'Scanlon's passing the buck . . .'

Indeed, they feel that Bough:

'was a lot more friendly towards Hugh Scanlon than he was towards Fraser.'
(a view shared by one of the most right wing groups, group 26)

and that Scanlon is evasive, rather than persuasive:

'He's pretty quick with his tongue, he can get out of any situation.'

They identify with the 'responsible' perspective preferred by Fraser, especially when he demands of Scanlon point blank that he 'tell us whether or not he wants another phase of pay restraint':

> 'The best question that was asked that I wanted to ask was that put by Ian Fraser . . .'

They validate the *Nationwide* team's claim to simply represent 'our' views:

> 'Bough was just picking up on the implications of what everyone was . . . asking in their own mind . . . I don't think it was him personally.'

and reduce the signifying mechanisms of the programme to a question of personalities:

> 'It was just a question of personalities rather than the way it was set up.'

in which the form of the treatment (an attack on a trade unionist) is naturalised/legitimated as 'given' by the nature of the situation (trades unions are responsible) and the role of the media in constructing this definition of the situation is obliterated:

> 'It's quite right, surely there's nothing that Fraser, try as he might, or any of the others, *can* do about the situation until he knows what Scanlon is going to do. I would say the weight, definitely, *is* on Scanlon . . .'

Group 20

An all male group of white, full-time, trade union officials (National Union of Hosiery and Knitwear Workers, National Union of Public Employees, Confederation of Health Service Employees) aged 29–47, with a working class background, on TUC training course; predominantly Labour.

This group inhabit a dominant/populist-inflected version of negotiated code, espousing a right-wing Labour perspective. They are regular *Nationwide* watchers and approve both the programme's mode of address and ideological problematic.

They take up and endorse *Nationwide*'s own 'programme values', they approve of the programme in so far as it is:

> 'Slightly lighthearted . . . takes it straight away . . . it's more spontaneous . . . as it happens, on that day . . . more personal . . .' 'I find that quite interesting . . . there's something in that programme for everyone to have a look at . . .'
> 'It seems to be a programme acceptable to the vast majority of people.'

To a large extent they do identify with the presenters and accept them as their 'enquiring representatives'. They remark of Bough that he was:

> 'basically saying what many of us thought last year . . .'
> 'Probably he was asking the questions millions of other people want to ask as well . . .'

The only fully oppositional element in this group's decoding comes in relation to the identifiably right-wing Conservative tax expert who bemoans the high level of public expenditure:

> 'You've got to look at the fact that in this country you've got services . . . hospitals, social security for those people who can't work or there is no work for them and it's got to be paid for. Now . . . if you've got that position . . . people like him who knock it should say what they would do . . .'

But it is a rejection of a *particular* position within an overall acceptance of the *Nationwide* framework – a classic structure of negotiated code (cf. Parkin 1973).

They feel that an unusual amount of 'space' is allowed to Mr. Worthington and that this is different from the kind of treatment *they* can expect from the media:

> 'I was surprised the interviewer never asked him other questions . . . a trade unionist would never get away with that . . . unless it was Jimmy Reid . . .'

They sympathise with the farm worker portrayed:

> 'Who is more emotive than the farm worker who does his best . . . but can't have his fags now . . . ?'
> '. . . anyone who lives in a rural area knows the problems of the agricultural worker . . .'

but feel that in this respect it is an open question whether the Chancellor has treated the farm worker unfairly, for, as far as they can see from the programme:

> 'The bloke that does the best out of it is the farm worker . . .'

They are critical of the middle manager's complaints:

> 'The point is you know, with us, we know, that middle management are not getting all that bad a deal . . . where they're not picking up in wages they're picking up in other areas . . . I mean let's face it . . . let's take the company car . . . I mean, he doesn't have his own car, "poor lad!" . . . they didn't . . . say that car's probably worth £20/30 per week . . .'

However, at the same time, they accept the individualistic, anti-tax theme of the programme:

> 'The only ideal situation *is* to have as much as you can to earn . . . and they find some way of taking it off you.'

and to this extent identify with the programme's construction of the plight of middle management, because if:

> 'we're talking about incentives . . . it's going to come to us as well . . .'

The 'we' constructed by the programme is something with which they do identify. Further, they accept the programme's construction of an undifferentiated national community which is suffering economic hardship:

> 'There's a sort of pathos in each of those cases . . .'
> 'It's not even the rich get richer and the poor get poorer . . . any more . . . it's *we* get poorer.'

Their involvement in right-wing union politics is evidenced by their acceptance that in the trade union/employer discussion Bough is simply representing our interests; and moreover that, much in the same way as the issue is presented by the programme, it is the unions, and not the employers, who are 'responsible' for the problem. They accept Fraser's self-presentation as a disinterested and objective onlooker who cannot 'do anything' himself:

> 'The things Fraser was saying . . . basically there's nothing sticking on what he says. . .'

They accept the programme's perspective on the unions, arguing that this is simply 'realistic':

> '. . . let's face it, it's the TUC that's going to make or break any kind of deal . . . basically what the interviewer was doing was saying, on behalf of you and me and everyone else in the country, are you going to play ball so we can have our [tax reduction] . . .'

and they accept the terms of the *Nationwide* problematic:

> 'I thought this was a programme that was fair. It was saying there isn't any incentive to try and advance . . . yourself.'

The sense of 'realism' goes further and is expressed as a Labour version of 'Realpolitik' unionism; this is how they characterise their own position as union officials:

> 'We're sitting here, in this place [TUC College] this week . . . learning basically to put other people out of work . . . this is basically what we do and when we boil it down this is what we do because we know that [for] a

vast majority of [our] productivity schemes . . . the only way is by putting people on the dole . . .'

Group 21

A group of white, mainly male bank managers, with an upper middle class background, on a two-week in-service training course at a private college run by the Midland bank; aged 29–52; predominantly Conservative.

The predominant focus of concern for this group is the mode of address or presentation of the programme. This is so out of key with the relatively academic/serious forms of discourse, in TV and the press, to which they are attuned that their experience is one of radical disjuncture at this level. This is a level of discourse with which they make no connections:

> 'Well, speaking for myself, if I'd wanted to find out about the Budget I'd probably rely on the next day's newspaper . . . something like the *Telegraph* . . . or the *Money Programme* . . .'

(From a quite different perspective, they repeat the comments of the predominantly black F.E. groups; cf. group 12: 'If I'd been watching at home I'd have switched off, honestly . . .')

Further, when asked:

> Q: 'How did that come across as a message about the budget?'

They replied:

> 'It wasn't sufficient, to be quite frank . . .'
> '. . . it didn't do anything for me . . .'
> '. . . I find that kind of plot embarrassing . . .'
> '. . . I just squirm in embarrassment for them.'
> '. . . I'd far rather have a discussion between three or four opposing views . . .'
> '. . . I mean it's much more rewarding . . . more ideas . . . they are articulate . . .'

It is *ideas*, not 'people', which are important to them:

> Q: 'What about the actuality sequences – going into people's homes?'
> 'I don't think you need it – if we're talking about ideas.'

Rather than the immediacy of 'seeing for yourself' someone's experiences, which many of the working class groups (e.g. 1–6) take as at least a partial definition of what 'good TV' is, for this group it has to be about considered judgements and facts:

> 'In that programme, what have we heard? We've heard opinions from various people which don't necessarily relate to facts . . . some of the

information . . . or background . . . all you've picked up are people's reactions . . . not considered . . .'
'I mean . . . the point was made [by Ian Fraser of Rolls Royce] "I'm not prepared to comment on the Budget till I've seen it in full tomorrow . . ." '

As far as this group are concerned, *Nationwide* are:

'exploiting raw emotion . . . they encourage it . . .'
'sensationalising items . . .'
'It's entertainment . . . raw entertainment value . . .'
'It's basically dishonest . . . I don't think it's representative . . .'
'As entertainment that's . . . maybe . . . acceptable . . . you can lead people by the nose . . . now if you're talking about communicating to the public and you're actually leading them, I think that's dishonest . . .'

In startling contrast, for example, with group 17's insistance that items should be short, fast and to the immediate point, a perspective from which *Nationwide* was seen to fall short, this group feel that *Nationwide:*

'. . . try and pack far too much into one particular programme . . . questions are asked, and before somebody had really got time to satisfactorily explain . . . it's into another question . . . and you lose the actual tack . . .'
'I can't bear it . . . I think it's awful . . . one thing . . . then chop, chop, you're onto the next thing.'

This concern with the coherence and development of an argument leads them to single out the interview with the tax expert, Mr. Worthington, as praiseworthy. They feel that the item was a little unbalanced:

'Particularly that accountant from Birmingham . . . was . . . very much taking a view very strongly, that normally would only be expressed with someone else on the other side of the table . . .'

But their predominant feeling is that at least the item contained a fully developed and coherent argument:

'There he was allowed to develop it . . .'

The programme certainly fails to provide this group with a point of identification, presumably because of the disjunction at the level of the programme's mode of address:

'I couldn't identify with any of them.'
'I didn't identify myself with the middle management . . .'

For them the whole tone of the programme makes it quite unacceptable to them, and they hypothesise, perhaps for others:

'There's a great danger, I'm sure Frank Bough isn't doing it deliberately, of being patronising or condescending . . . and this I found irritating – that "there's going to be £1.20 on your kind of income" . . . to me, Frank Bough on £20,000 a year . . . it's enough for a . . .'

They hypothesise that the target audience is:

'The car worker . . . the middle people . . . and below.'

and wonder aloud that the programme might have been:

'talking down . . . even to the lowest paid worker.'
'They place an emphasis on what this meant to the British worker . . . to a range of workers . . . I think it needed the same thing in a much more intelligent way – appealing to the more intelligent aspects of the people involved . . .'
'I wonder if they've underestimated their audience.'

But this is a perspective which is not unchallenged; their view of the 'middle people . . . and below' also leads them to wonder:

'Would many of the population be capable of absorbing the information . . . even the simple part of the question . . . especially in a programme of that sort . . . ?'

Because, they argue:

'They do not understand – the man in the street does not understand the issues: they understand "£10 a week" . . .'

The ideological problematic embedded in the programme provokes little comment. It is largely invisible to them because it is so closely equivalent to their own view. The lack of comment is I suggest evidence of the non-controversial/shared nature of the problematic. Indeed, they go so far as to deny the presence of *any* ideological framework; it's so 'obvious' as to be invisible:

Q: 'What was the implicit framework?'
'I don't think they had one . . .'
' . . . there wasn't a theme . . . like an outline of the budget . . .'

The only point made by the presentation of the budget, as far as they can see, is:

'It left you with a view . . . the lasting impression was that [Healey] didn't do very much for anyone . . .'

But they are very critical of this 'superficial' view precisely because it has not explored what they see as the crucial socio-political background:

'There is another side to the coin, he didn't do a lot, but there was full reason at the time why he couldn't do a lot, and that was virtually ignored . . .'

Group 22

An all male group of white, full-time, trade union officials (Transport and General Workers Union, Union of Shop, Distributive and Allied Workers, Union of Construction, Allied Trades and Technicians, Bakers, Food and Allied Workers Union, National Union of Agricultural and Allied Workers), aged 29–64, with a working class background, on a TUC training course; exclusively Labour.

The group find the problematic of the programme quite unacceptable and accompany the viewing of the videotape with their own spontaneous commentary:

Programme	*Commentary*
Link after vox pop interviews: 'Well, there we are, most people seem agreed the tax system is too severe . . .'	'That's a bloody sweeping statement, isn't it? . . . from four bloody edited interviews!'
Interview with Mr. Worthington:	'Is this chap a tax expert? Seems like a berk . . .' 'Poor old middle management!' 'Aaaagh!'
'. . . ambitious people . . .'	'Avaricious people, did he say?' 'What about the workers!!' 'Let's watch *Crossroads.*'
'. . . and of course the lower paid workers will benefit . . . from . . . er . . . the . . . er . . .'	'Extra crumbs falling from the table!'
Three Families Section–Manager: 'He doesn't own a car . . .'	'Ha, Ha! That's a good one!'
'A modest bungalow . . .'	'Family mansion! His lav's bigger than my lounge!'
'. . . we can't have avocadoes any more . . .'	'We had to let the maid go!' 'Did you hear that!'
Mr. Tufnall digging in his garden	'Those aren't Marks & Spencer's shoes he's wearing.'
'However much you get– someone else is waiting to take it away . . .'	'Good! Redistribution of wealth and fat . . .'
'What, of course is a tragedy is in respect of his child still at college . . .'	'They didn't mention that for the other peasants.'

'One has to run a car . . .'
'Does one! He doesn't "run a car".'
'His child at college . . .'
'I worked nights to do that.'
'actually in the hands of the trades unions . . .'
'Yurgh!'

This group began by commenting that the programme was:

'Obviously contrived, wasn't it, the whole thing . . . all contrived from start to finish to put the image over . . . I'm of the opinion the ones we've got to watch for the image creation are the local programmes.'

This they see as an unacceptably right-wing perspective, also seen as characteristic of:

'most ordinary TV programmes; serials . . . you get, em, *General Hospital*, it's so right-wing it's unbelievable – it's pushing the senior management at the people all the time, "you must respect the consultants and doctors", and "they're the people who make the decisions . . . and they know what they're doing". . .'

They say of the vox pop sequence in the programme that it is far too narrow and class-specific a sample of opinion to provide the 'ground' which *Nationwide* represents it as providing for their 'summation' of 'what most people think'.

'Then the way the actual interviews were . . . very carefully selected in the centre of Birmingham, mid-afternoon . . . with the shoppers and businessmen – there wasn't one dustman around . . . there wasn't any agricultural workers with their welly-boots on . . . it was purely middle class shoppers out buying their avocado pears or something; then at the end he says "everybody agrees" – he's met four people. I don't know how many people live in Birmingham, but there's more than four . . . he only shows what he wants to show.'

Mr. Worthington, the tax expert, is dismissed as a 'berk':

'Of course everybody believes that this chap is the expert, the TV tells us so, and the things he was saying, he might as well be reading a brief from Tory Central Office, which I think he probably was anyway . . . and they didn't just ask him to lay out the facts, they actually asked him his opinion, and to me an independent expert is supposed to tell you the facts, not necessarily give you their opinion on general policy . . .'

The group feel that Mr. Worthington is allowed 'free rein' in the programme, very different from their experience of being interviewed by the media:

'The development of the scene was allowed to go on and on, wasn't it? Glamour boy [i.e. the interviewer] just sat back and let him get on with it . . .'

'*We've* found that local sort of media – y'know we've got good relations – and yet we're cut all the time, as compared with the management's views.'

The group do, at one point, comment on the form of the programme's presentation, or mode of address:

'My major complaint against most of the *Nationwide* programmes, apart from the political ones, is the way in which they trivialise every topic they seem to take up – and just when the topic begins to blossom out, they suddenly say, "well that's it. . ." '

But crucially, as distinct for instance from group 21, for whom the mode of address of the programme is the dominant issue, for this group it is the political perspective or problematic of the programme which is the dominant focus of their concerns. The perspective is one which they vehemently reject:

'The perspective was that of the poor hard-pressed managerial section . . . they had the farm worker there . . . that was, sort of, "well, you've got £1.90 now – are you happy with that – now go away" and then "Now, you, poor sod, you're on £13,000 p.a. . . . and a free car . . . Christ, they've only given you £1.10 – I bet you're speechless!" . . .'
'. . . the sympathy was, you're poor, and you're badly paid and we all know that, because probably it's all your own bloody fault anyway – the whole programme started from the premise that whatever the budget did it would not benefit the country unless middle management was given a hefty increase – that was the main premise of the programme, they started with that . . . they throw the farm worker in simply for balance at the other end of the scale. . .'

The visibility of this distinct 'premise' for this group is in striking contrast, again, with group 21, for whom this premise is so common-sensical as to be invisible and for whom the programme had no particular theme or premise of this kind.

This group feel that *Nationwide,* because of its politics, is not a programme for them. It is:

'Not for TU officials. For the middle class. . .'
'Undoubtedly what they regard as being the backbone of the country, the middle class . . . they allowed the agricultural worker to come in so as the middle class can look down on him and say "poor sod, but I can't afford to give him anything because I've had to do without me second car, etc!" '

As far as they are concerned the whole union/employer discussion is totally biased against Scanlon:

'He [interviewer] was pushing him into a corner . . . that was the first

comment, immediately getting him into a corner, then the opponent [i.e. Fraser] who was supposed to have been equal . . . more or less came in behind Bough to support Bough's attack on Scanlon.'
'Yes, except you've got to realise Scanlon slipped most of those punches expertly – a past master . . .'
'but pointed, direct questions . . .'

There is, however, another thread to this group's comments which emerges particularly around the question of tax and incentives. In line with the group's dominant political perspective, there is some defence of progressive taxation:

'What about the social wage? It's only distribution . . . the taxation takes it from you and gives to me . . . I mean, if you're not taking income tax from those who can afford to pay, you can't give anything to those who aren't paying. . .'
'So long as I get benefit for the tax I pay, I'm happy enough.'

But at this point a more 'negotiated' perspective appears which shares much in common with the Labour 'Realpolitik' of some of group 20's comments; they say that to criticise *Nationwide*'s perspective on tax and incentives is misguided:

'It's not necessarily a criticism of the programme . . . a lot of highly paid skilled operatives fall into exactly the same trap . . . they probably listen to the programme themselves . . .'

As they somewhat uneasily put it, extending in a sense, some of Scanlon's own comments on the need to 'look after' the 'powerful . . . skilled elements':

'One of the main objects of the full-time official is to maintain differentials . . .'
'I'm not saying differentials are good . . . but as a trade unionist you've got to be able to maintain it. . .'

Indeed they extend this to a partial defence of *Nationwide*'s perspective; at least to an implicit agreement about what, in matters of tax, is 'reasonable':

'. . . I think we should try to get the income tax in this country to a respectable level to allow everyone to work and get something out of it . . . because there's no doubt about it, the higher up the ladder you go, the harder you're hit for income tax.'

They remark in justification of this perspective that the problem is that 'incentives' have been destroyed, because:

'They've increased the income tax in this country to such a degree that it don't matter how hard you work. . .'

Indeed, they also take up one of the other themes of the dominant media discourse about trade unionism:

> 'There's a lot of . . . unions in this country that could produce a lot more . . . British Leyland's one for a start-off.'

This is not to say that they wholeheartedly endorse this negotiated/'realist' perspective. They cannot, for it is in contradiction with much of the rest of their overall political outlook. It is rather to point to the extent to which this is a discourse of 'negotiated' code, crossed by contradictions with different perspectives in dominance in relation to different areas, or levels, of discussion.

Group 23

A group of mainly white, male and female shop stewards and trade union activists (Transport and General Workers Union, Civil and Public Services Association, National Society of Operative Printers, Graphical and Media Personnel, Society of Graphical and Allied Trades), with upper working class backgrounds, aged 23–40, studying part-time for a diploma in Labour Studies at a London Polytechnic; predominantly socialist or Labour.

Like group 22 they produce a spontaneous 'commentary' on viewing the programme:

Programme	*Commentary*
Mr. Worthington	'Aah, poor middle management.' 'Who does he represent?' 'Aargh.'
'few very good hard workers'	'Oh, no . . .'
'benefit from the extra . . . er'	[interrupts] 'Exploitation!' (cf. group 22) 'As if everyone's aspiring to be middle management!'
Dallason: 'not at expense of services'	'Good lad'
Bough: 'every little does help . . .'	'Fucking bastard!'
Manager: 'he doesn't own a car . . .'	'Tell us another.'
Mrs. Tufnall	'Listen to that accent! Dear me . . .' 'Yeah, you really starve on £7,000 p.a.!'
'. . . speechless is the right word . . .'	'Shut up then!'
'One has to run a car . . .'	'It's paid for by the company!'
'. . . he's doing chartered accountancy'	'Learning the tax fiddling game!'
'We're actually in the hands of the unions.'	'Here we go!'

Like many of the other working class groups they express a preference for the less reverent and staid style of the ITV equivalent to *Nationwide*, *London Today*:

> 'It's amazing, the difference in *Nationwide* and *Today* – the BBC just happens to be quite a lot more patronising.'
> 'There used to be a Friday night debate on *Today*, sort of open forum, unrehearsed . . . and you'd get ordinary leaders of the public, spokesmen and they would be totally unrehearsed, so if it was say, the leaders of the GLC they would come under a lot of genuine fire that you would be able to see . . . members of the audience getting up at random and saying what they've got to say you know, I used to think that debate was quite reasonable.'

They can to some extent endorse *Nationwide*'s mode of address or at least not dismiss it by reference to the criteria of 'serious current affairs'. They grant that *Nationwide* is:

> 'light entertainment isn't it? I mean, you know, there is that . . .'
> 'It takes the issues of the day and it is quite entertaining.'
> 'It was easy watching.'
> 'You know, like, did you see that guy trying to fight that budgie or something, and he lost! . . .'

However, they object, on the whole, to *Nationwide*'s:

> 'sort of soothing, jolly approach . . . as if you can take a nasty problem and just wrap it up . . . you know – "we're all in the same boat together" and clearly we're all going to live to fight another day . . . you get rational, cheerful people . . . cheering us all up just a little bit at the end so it's not so bad. . .'

But the 'style' of the programme is not their overall concern; even a comment that begins by focusing on this aspect shifts in mid-stream towards their dominant focus of interest, the class politics of the programme:

> 'I'm not discussing the actual credibility of the arguments used . . . but it's the sort of jolly show-like atmosphere they create, and there's all these people laughing at their own misfortunes . . . as if they [producers] would encourage that resignation . . .'
> '. . . except for middle management – and they weren't encouraged to do that then – it was "you must be speechless", and there's no jollity there – it's "tragedy" . . . that very sympathetic "umpire" said to this guy . . . he used the word at least four times, he said "unfortunately" at least four times in a minute, which is amazing.'

They reject what they see as the programme's highly partial treatment of the personnel manager:

'I didn't believe it anyway. I don't believe the personnel manager at Tate and Lyle's only gets that.'
'Yeah, but he's got a car . . . benefits . . . the £7,000 is the declared tax value, the rest comes in benefits.'
'And they had more sympathy with him . . . they used the word "unfortunate" . . . "unfortunate" all the while with him . . .'

Similarly they see the trade union/employer discussion as highly partial:

'Even in BBC terms, there wasn't any neutrality in it . . . at all.'
'The guy from Rolls Royce didn't even have to open his mouth. I was amazed, he just sort of said "We can ask Mr. Scanlon and he can hang himself".'
(cf. those groups who interpret the fact that Fraser *talks* less to mean that the programme is biased against him.)
'He said to Scanlon, "well look, the burden is with you . . . you're fixing his wages" . . . union leaders are always being told, "well, you're running the country".'
'And then when he said to that bloke, "well you must be shattered. Industry must be depressed." (That's right!) ". . . we're all in the hands" (That's it!) "of the unionists . . ." (The link's there all the time) yes, as you've said the debate was confined in a very narrow channel . . . you see this is the problem . . .'

Although, in a rather cynically well-informed way, they feel that Scanlon 'did all right':

'Actually, Scanlon was very crafty, all that "rank and file" stuff . . . "my rank and file will decide" – his rank and file never knew about the pay policy till it was made compulsory . . .'

They reject the programme's claim to represent 'us':

'What was really frightening was the first line of the whole programme which was *telling* us what our "grouse" was, the main grouse, isn't it, you know is Income Tax.'

Further, they reject what they see as the programme's premise of national unity of interest:

'The whole thing about that programme was that really there's a unity of interest, I mean . . .'
'Workers who talk about tax being too high/lack of incentive – which is the sort of thing I remember the Tories pushing (Right) and the media hammering like "no-one wants to work" . . . poor sods at various levels of the social scale all hammered by exactly the same problems . . .'
'When they said, "well, how are they coping?" they had the poor old middle management, er, digging in the garden . . .'

'They've got these bits in common – that they'd all be prepared to do a bit of digging if they had to, you know, all of them, you know. Yeah, it's rather like when the war's on, "all ranks, er, go and do their bit," you know.'

They do not identify with the 'we' which they see the programme as attempting to construct:

'I mean they want 'we' and they want the viewer, the average viewer, to all think . . . "we" . . . That's why we've got these three families (Yeah) and they appear so sympathetic so that the "we" sticks, "we are" . . .'

The group are highly critical of what they see as a whole range of absences within the programme, issues excluded by *Nationwide*'s problematic:

'There's no discussion of investment, growth, production, creation of employment . . .'
'Well, nobody mentioned unemployment. I mean that's pretty amazing . . .'
'. . . the whole, I mean . . . budgets in the past have always been to do with the level of employment, to get through the whole thing without even mentioning the fact that it's not doing anything about unemployment!'
'There's no real, I mean, not an intelligent discussion of economics, actually. You know, no talk of, em, em, investment, er, productivity, how to expand the economy, anything like that at all . . .'
'. . . as if all the economy hinged on taxation control or, or . . . (Yeah) There's no such thing as growth or, or, investments.'

In particular they are critical of what they see as the programme's 'one-sided' approach to the problem of tax:

'The line was . . . we haven't had any tax cuts, and that's what's wrong with the country.'
'And the whole thing. There was no mention at all about what happens to taxes. Should we be taxed, you know. The only answer is, Yeah, of course we should be taxed.'
'There's no mention anywhere in all this run up against "taxation's bad" and "you shouldn't have to pay this sort of level", that there are those people that, through no fault of their own, are so lowly paid that they have to have it bumped up by somewhere else, which has to come out of the tax system . . .'
'Yeah . . . tax is a drag . . . but like it's a drag to have to drive down the left hand side of the road! – I mean, trying driving down the other side!'

They are aware of odd moments in the programme in which they feel these issues begin to surface but feel that they are quickly dismissed:

'There was one guy [Dallason] made one very interesting point, that . . .

> yes, we need, incentives all right – but not at the expense of cutting hospitals and schools – and that was never taken up.'

Similarly they feel:

> 'There was one brief mention of pay policy. Scanlon picked that up at the end but that was in the last two minutes and he's a big bad trade unionist anyway.'

And crucially:

> 'No reference whatsoever to stocks and shares. You know, the things that are accumulating money all the time without lifting a finger, you know, no reference at all . . .'

The group's perspective is oppositional in the fullest sense: they not simply *reject* the content or 'bias' of the particular items – they *redefine* the programme's problematic and implied evaluations into quite other terms:

> 'And always the assumption that people over a certain level . . . in a certain job have got a basic right to eat avocado pear and everybody else hasn't, for example . . . I mean we might put the boot on the other foot and say somebody who does manual, physical work, coupled with his head as well, so that he doesn't crash into trees with his tractor, em, you know, would need to eat equally well, if not better (Right), since the middle class don't.'

They feel precisely that *Nationwide* is:

> 'supposing that that emphasis should be put . . . you know, as if it's more important the fact that you're a Chartered Accountant, more important than somebody who's actually digging the food that we eat . . .'

As far as they are concerned the relative value and importance of the different categories of labour represented in the programme is quite different from that suggested there:

> '. . . there's this big emphasis by almost everybody, the presenters, the experts, apart from individual workers, about middle management. (Yeah) Now this runs all the way through. And right at the very end you suddenly get the fact that it's Scanlon and all these, you know, terrible union people that are ruining industry because they keep asking for too much money. (Yeah) It was amazing you know, I mean there's a guy who is a farm worker that produces the stuff that we eat. Nobody thinks *he's* essential. There's a skilled tool room worker, nobody thinks *he's* essential . . . there's that personnel manager, which for all our experience, you know [laughter] they could hang up tomorrow and nobody'd notice the difference, *he's* suddenly essential, extremely important to the whole of industry, i.e. the economy, i.e. the nation (Right) . . . it's a political line that runs under the whole thing.'

This implicit theory simply does not fit with their own experience:

'The whole point was, it would appear to be that the farm labourer is irrelevant, the toolroom worker's irrelevant, everybody's irrelevant barring the manager . . . one little strata within industry that is so important . . . where, I mean, all our experience is that they do more bloody harm than good. Get rid of them – when this Ford's thing was going on recently, when the foremen came out on strike for two days, and they had to scuttle back because it didn't make a blind bit of difference . . . suddenly people had found out, they didn't need them any more . . .'

Moreover, it is not simply a rejection at the level of 'experience'; it is also a theoretical rejection, in terms of an economic theory about the origins of value, opposed to the theory of classical economics which they see enshrined in the programme:

'And this is the belief in the entrepreneur's special skill which makes wealth appear like magic (Yeah) by telling all these idiots what to do, you know, er, it's a sort of special skill. Well, yeah, it's, mind you it really relates . . . to classical economic theory, the point there is that you see the factors of production, of inputs, workers . . . costing and everything else . . . and it's only the sort of skill of the overall managers and all their executives, who can sort of cream off, you know, this exact pool of skill and machinery and get profit from somewhere and therefore these individuals, the ones who create profit because it's their judgement and skills who produce it, not the actual graft, the workers, who should get surplus, then take home a certain amount, you know, that's obviously . . . it's two totally different interpretations of where wealth comes from – basic stuff.'

The group have reservations about the representation of women in the programme too:

'Interesting the way they use women too: "I ask Kathy about the family budget", like she stays at home and looks after the kids and feeds the cat.'

Moreover, they do not simply reject this programme, or *Nationwide* as such:

'The thing is, I mean, I don't think you can really, sort of, take *Nationwide* in isolation, I mean take *Nationwide* and add the *Sun*, the *Mirror*, and the *Daily Express* to it, it's all the same – whole heap of crap . . .'
'. . . then you begin to realise, you know . . . the whole thing is colouring everything.'

Interestingly, it is also this group who are most articulate about the social/political conditions of their 'reading' of the programme, seeing it clearly in terms of socio-political structure rather than simply being a question of different 'individual' views:

> 'Let's be fair . . . and say we watch that as a group of people very committed to a certain thing . . . I mean, no doubt your reactions for a bank manager would have been totally different.'

Group 24

Owing to a fault in the tape recording this group had later to be omitted from the analysis.

Group 25

This is a group of mainly black students (West Indian, Zimbabwean) mostly women, ranging in age from 18 to 37. The group has a lower working class background and is in full-time further education studying 'A' level sociology at an F.E. college in Hackney; predominantly Labour.

The group's response to *Nationwide* is overwhelmingly negative:

> 'I go to sleep when things like that are on . . . it's boring.'

And this is partly to do with their lack of identification with the image of the audience inscribed within *Nationwide*. They are clear that it is not a programme for them, but for:

> 'Affluent people . . . middle class people . . . When they were talking about tax forms [i.e. in the vox pop section where the audience is presented with a surrogate image of itself], all those people seemed fairly middle class.'

– almost exactly repeating the comment of the TU group 22: 'all those middle class shoppers and businessmen . . .'

When asked to describe how they saw *Nationwide*'s basic perspective they again identified it as one they rejected, moving from an account of this rejection within the terms of their sociological education to one in terms of their own viewing:

> '*Nationwide*'s Conservative. Always within the dominant normative order . . . Recently *Nationwide* have been very conservative – they did a thing on Astrid Proll – they made those prisons in Germany look very, very comfortable and everything, you can walk around wherever you want to go . . . They made the people who supported her seem very, sort of, into their own political games, living in communes and that.'

The programmes that they do prefer, and identify with to some extent, are those that are more directly grounded in some localised sense of working class life, in whatever limited and mediated ways.

They prefer the ITV equivalent to *Nationwide*:

'There's plenty of pictures and a lot less talk . . . perhaps it's because it's local, it's not nationwide . . . perhaps it's more in touch with its audience . . .'

And they (surprisingly at one level, given the disjunction between their black sub-culture and the white/northern culture of the programme – though they share a sense of class) identify with:

'*Coronation Street* . . . perhaps [laughter]. Well, maybe . . . we hear people coming in and saying "Did you hear what happened, it's like mine . . . ", you know, and people do tend to identify a bit with that. Not that they say, "Well, that's an exact replica" – of course it isn't, but you know, like gossip corners in the local pub, and the rest of it . . .'

But beyond these mediated identifications there is also that very small segment of broadcasting which this, predominantly black, group can relate to directly, and which also raises resentments in so far as they feel that it is ignored by the white audience:

'I remember when *Roots* was on, and I was dashing to watch it, sit glued to the TV and I might sit and discuss *Roots* and then I'd go to work and nobody'd say a word about it. And if you do, eventually, "Yeah, it's a good programme" – and shut up. When *Holocaust* was on, everybody discussed that.'
'. . . it gets on to *Roots* and *Martin Luther King* and you get a block. I mean you could talk about it on your own. You could talk about it among your friends and your relatives because they could identify with it as well as you, but if you . . . I'm not saying all . . . but a lot of white people . . . they don't want to know. They say, "yes, it's a good programme", or, "yes, I can see why King had to", you know, what they were fighting against, but no, they don't want to know.'
'They won't see it's anything to do with them.'
'They don't want to know so much . . .'

The group make a consistently oppositional reading of *Nationwide*, basically focusing on, and rejecting wholesale, its problematic; by contrast, the mode of address is not commented on. They reject the programme's 'hierarchy of credibility' as between different speakers because of their class perspective:

'. . . in *class*, I would have more sympathy with the agricultural worker than I would with the management person, much more.'

although they can see that the discourse of the programme (and its preferring of certain codes of speech and presentation taken over from other social/educational hierarchies) privileges the manager; because:

'the farm labourer, from his own experience, the way he put it over, I mean he wouldn't know how to put it over as well as the manager, would he?'

They accept that their reading is in disjunction with that preferred by the programme, and they reject the preferred reading precisely because of its lack of any class perspective:

> Q: 'Does your reading fit with what the programme is doing?'
> 'No, I just disagree – no, I don't think it does. I think the programme's just more or less saying, [i.e. at the level of connotation, meta-language] "look, it's not just the working class, it's not just the lower class people who are being treated like this, it's all the classes." I think that's all it's doing, I don't think it's really looking at it from any real [= class?] point of view.'

The tax expert, Mr. Worthington, is decoded in a way that precisely reverses the 'generalising mechanism' of universalisation – from particular to general interest – which is embedded in his speech. His speech is deconstructed quite crudely to the level of self-interest:

> 'You got a feeling it would obviously work *for him* . . . he said that it would benefit directly middle management and increase incentive, and indirectly benefit the working population [cf. the TU groups' interjections in his speech here] . . . because the middle management would be better off, more gross product would be raised . . . he was directly for the incentive approach, the top . . .'

This reading is in striking contrast with that of the trainee print management group 26, who argue that he is not talking about 'his personal salary . . . it's more profits in order to reinvest . . . and create new jobs . . .', obviously reading the speech at the level of the 'general interest'.

Again in contrast with group 26 (who see the 'trade union' discussion as biased in Scanlon's favour) this group clearly read this discussion as structured around a problematic which is anti-union in a way they totally reject. The Rolls Royce chairman and the presenter are seen as uniting powerfully against Scanlon:

> 'They kept going on about him, the union running the country . . . I could see they were goading Hugh Scanlon . . . focusing on him, and this other manager . . . chairman of Rolls Royce, like sitting back cool . . . like sitting back "with us", saying, "look at what he's doing to us, and I'm also included . . ." '

But they also think that, despite this structure, Scanlon is not 'caught out':

> 'I don't think so . . . some of the questions he was very clever at answering, especially the last one . . . where they were trying to make him

say what . . . and he said, "well, I'll take it to the unions and see what they say, and I'll have to, you know, act on that." I thought . . . he really got out of that one well.'

The group read the whole 'Families' section as clearly weighted in favour of the personnel manager:

'I thought the middle management guy got free rein in a way – 'cos they just asked a question and he went on and on . . .'
'They brought him over in a really good light . . . he's seen as desirable . . . necessary to the future of the company . . . so he should be in a much better position – yeah: when his £1.10 popped up, you know – "GASP, SHOCK, HORROR" . . . it came over as if he was badly off – you know, £1.10 – gosh!'
'Will you be able to feed your kids?!'
'I mean, they're *really* hard done by!'

And they are particularly critical of the absences in the programme's perspective on the manager:

'They didn't say anything about the manager's entertainment . . . the way the managers entertain themselves, his leisure, they didn't say anything about that at all – I mean, apart from her saying, "oh, we won't have any more avocado pears!" . . . being a manager you expect him to entertain a bit, and they didn't say anything about that at all, what they spend on their leisure time . . . and the fact that he ran a company car as well, and they pay.'

It is the moment when the manager's wife complains that she is unable any longer to afford avocado pears which reduces them, beyond mere criticism, to incredulous laughter:

'It was just the way she said [i.e. in a very 'plummy' accent], "well now, we'll have to cut down on our eating!" It didn't look like they had at all, you know! . . . that's a *luxury* she's talking about – she said you can't have avocado pears!!'

Moreover, they are aware of another level of poverty off the bottom of the scale shown in the programme:

'. . . another thing . . . they showed the agricultural labourer, they didn't show the labourer in the industrial city . . . the agricultural labourer . . . OK, he earns such and such, but he's able to have his little bit of land to grow his vegetables . . . he can supplement his income, and the industrial worker can't . . .'

More than any of the others, this group also produced a specifically feminist critique of the 'Three Families' section in its portrayal of sex roles:

'When they were talking about it . . . they said to her, "how does Ken's money go?" They didn't ask her about her money, what she did with her money. As far as they were concerned that's pin money . . . just like when they talked about the agricultural labourer – they talked about his luxury, what was *his* luxury, but not hers, which I thought was really typical.'

Moreover, to take up the point made by MacCabe (1978), ideologies do have to function as 'descriptions' or 'explanations' of the 'reader's life', and it is in so far as they succeed or fail to do this that 'the ideological viewpoint of the text' is accepted or rejected. This group reject the 'description' of their life offered by the programme. They can find no 'point of identification' within this discourse about families, into which, Bough has assured us, 'most people in Britain' should fit. Their particular experience of family structures among a black, working class, inner city community is simply not accounted for. The picture of family life is as inappropriate to them as that offered in a 'Peter and Jane' reading scheme:

'It didn't show one parent families, how they would count, the average family in a council estate . . . all these people seemed to have cars, their own home . . . property.'

Further (and this is to validate John O. Thompson's (1979) critical remarks on *Everyday Television: 'Nationwide'*) which argued for a reading of '*Nationwide*'s discourse as one which is no less in flight from Freud than from Marx') this group are very aware of the inadequacies of *Nationwide*'s portrayal of family life, in terms of the absence of any idea that the whole domestic field is also one of struggle and contradiction:

'They show it . . . like all the husbands and wives pitching in to cope with it, cope with looking at problems. I mean, they don't show conflict, fighting, things that we know happen. I mean it's just not, to me it's just not a true picture – it's too harmonious, artificial.'

Or, put more simply:

'Don't they think of the average family?'

which of course *Nationwide* would claim precisely to do. This simply shows the lack of a 'unitary sign community' in relation to central signifiers like 'family'. The representation of 'the family' within the discourse of *Nationwide* has no purchase on the representation of that field within the discourse and experience of this group – and is consequently rejected.

Group 26

A small group of male, mainly white, European management trainees, between the ages of 22 and 39, with a principally upper middle class background; predominantly Conservative.

The basic orientation of the group towards the programme is one of criticism and disjunction, but from a trenchantly right-wing Conservative perspective, which sees *Nationwide* as being to the political 'left'. This may be the result of the lack of fit between the respective political spectrums of the speakers' countries of origin – South Africa, Belgium, Greece – and those of Great Britain; however, this political disjunction co-exists with a simple lack of familiarity with the codes of British television and their dominant political connotations.

Where group 25 criticised the programme for the lack of any 'real' (class?) perspective this group find that aspect of the programme's problematic quite acceptable and in keeping with their own perspective:

> 'I think it's pretty evenly balanced, they say, "well, this guy isn't better off, the lower paid aren't better off," you know, it's really all the same thing, nobody's really better off.'

However, the international scales of comparison, when it comes to income figures, introduce complications:

> 'They regarded that personnel manager as already a high wage – it's not a high wage, £7,000!'

From this point on a major theme in the discussion is that *Nationwide* is seen as a left-wing programme:

> 'An awful fact of *Nationwide* is that they're very subjective; the people are very pro-Labour . . . I watch them very often, they're always biased. If there's anybody, for example, who earns £10,000 p.a. they are sort of looking at them as if they were a rich bastard. That's what I've always found on *Nationwide* . . .' 'They mentioned something . . . the person who has more than £20,000. Then they were saying, "oh, yes, the bastard . . . " I find that extremely biased, saying something like that . . . I watch them very often, *Nationwide,* that's why I have this thing in my head, it's always the same thing with them . . .'

From their perspective this group decode *Nationwide*'s uneasy combination of 'radical populism' as simply radical. In a sense this group do not share the cultural codes within which the discourse of *Nationwide* is situated – to the extent that they even interpret the programme as being against small businesses:

> 'I've always found that *Nationwide,* they rank people, as soon as they have a bit too much money, like this thing pidgeonhole, they really drag people through the mud, because they're small businessmen, they're not regarded as very important.'

Further, their decoding of the trade union/employer discussion is focused on the form only – who talks most – rather than on the structured dominance of the employers' perspective, which is not a relation of force established on the basis of who gets most speaking time, but rather of whose problematic sets the terms of the discussion:

'The guy from the union said everything, then they ask something from the man from Rolls Royce and immediately the guy from the union had the last word again.'
'. . . they didn't give him a chance, the guy from management.'

The 'explanation' of *Nationwide*'s supposed political radicalism is later produced through a tightly economistic perspective:

'I mean it's BBC and ITV. ITV can't be socialists because it's private enterprise. BBC is a state-owned thing, so it's socialist. That's how it works in this country.'

– a perspective which will clearly have no truck with the quirks of relative autonomy!

However, there is also a subordinate thread of interpretation in the group which goes somewhat against the grain of this trenchantly conservative view. The students' immersion in an academic discourse of business economics (as trainee managers) leaves them dissatisfied with the level of complexity of *Nationwide*'s arguments:

'You can't talk about taxation levels without saying what public expenditure is about . . . I mean higher tax may mean you have a better post office . . . road communications . . . it's not really a decrease in your income, it's just another way of seeing your income. If you don't see the figures of public expenditure you can't really speak about higher or lower incomes.'

and again:

'I find it's a silly idea – very naive to say you can lower public spending . . . I mean you can say cut public expenditure, but in what ways? Is it defence, is it transport, what is it? Housing?'

And at this point one of the more reticent speakers moves the discussion from the level of criticism of the programme's lack of complex argument towards the level of political criticism, from a Labourist perspective:

'The part of the system I really don't understand or support is the supplementary benefits they give to people. I think people who work should be given enough basic salary that they would not have to go for supplementary benefits – it's not fair enough, because a lot of people are ashamed to go in for supplementary benefits . . .'

– a problematic clearly to the left of those considered by *Nationwide*.

However, this contribution is out of key with the views expressed by the other members of the group and is not developed.

The dominant conservative theme is also articulated with the 'academically' orientated criticism that *Nationwide* fails to develop sufficiently complex arguments. Surprisingly, this is even a point made about the interview with Mr. Worthington (the tax expert) who was identified by many of the groups as being allowed free rein to develop his points:

> 'That's always the case in *Nationwide*, they try to make you say whatever they want you to say. They don't allow you the time, or give you the chance to explain what you think and go deeper.'

These criticisms of the *form* of this interview notwithstanding, the perspective of classical free-market economics which Worthington is advancing is heartily endorsed by the group. As noted, group 25 decode this as a truly 'ideological' speech, disguising personal/sectional interests as representative of the universal/general interest. Group 26 defend Worthington's argument in a precisely contrary fashion:

> 'It's not him personally . . . he does not say the manager should be paid more – he says that business should be allowed to have more profits in order to reinvest, and it circulates money and creates new jobs . . . makes money go round . . . so I'm not sure he was talking about salaries or wages: he was talking about expenditure and investment.'

And, to the suggestion that an increase in profits will not necessarily lead to an increase in investment, their developed free-market problematic of course replies:

> 'That's because management in this country is afraid of investing. They've made management afraid of investing because it's a very uncertain time.'
> Q: 'Who's made management afraid?'
> 'The TUC, the general appearance of the government. Everything seems to be working for Labour . . . but in the wrong way, however . . . they don't look into the future . . . to tomorrow's situation . . .'

The problematic is internally coherent and far-reaching, moving from the justification of personal response:

> 'I come from a very conservative family. Several times I've wanted to pick up the phone and phone *Nationwide*, because I have seen people being pulled through the mud there because they have too much money.'
> '. . . *Nationwide*, for them, they're pigs, the pigs of this society who rob all the money.'

to a fully blown defence of the classical economic theory of the origin of value (cf. the trade union groups' insertion of the labour theory of value here):

> 'But you must not forget these are the people who provide this country with everything this country has . . . they are the ones who are employing the people . . .'
> 'Of course they deserve it. People are not paid according to what they can do here in this country.'

and in conclusion makes the emphatic point:

> 'And if you have a free market system, everybody is free to do what he wants . . .'
> '. . . if you want to get something, you can do it.'

Note
In a sense this group are so far to the right of the political spectrum that they might be said to be making a right-wing 'oppositional reading' of *Nationwide*, which they take to be a 'left'/'socialist' programme. However, the substance of their arguments pulls in precisely the same direction as those of the discourse of *Nationwide*; apart from the difficulties of international comparisons and unfamiliarities, there is perhaps more similarity between their perspective and that of *Nationwide* than they realise. Considered only in a formal sense – i.e. in the form of the relation between the group as subject in relation to the signifying chain of the discourse – the responses of groups 25 and 26 could be considered to be equivalent: both are 'subjects out of position' in relation to this signifying system. However, the inadequacy of such a formalistic approach is precisely revealed by the way in which this formal equivalence is displaced through the quite different political/ideological problematics of the two groups. It is the articulation of the formal *and* ideological levels of the discourse, and its decoding, which is in question.

Group 27

This is a group of 18/19 year old white, male, upper working class apprentice printers, members of the National Graphical Association; no dominant political orientation.

They, like group 26, characterise *Nationwide* as:

> 'To the left, they're going to the left . . . the majority of people think that *Nationwide*'s left . . .'

and again, like group 26, concentrating on the form of the union/employer discussion rather than the structure of it, read it as biased in the unions' favour:

'They're talking more to Hugh Scanlon, aren't they?'
'More questions to him . . .'

They are clearly unhappy with what they take to be *Nationwide*'s pro-union stance. Their own experience of the print unions is predominantly one of cynicism and self-interest disguised as principle. From this perspective they are cynical about the presentation of the case of Ken Dallason, the Leyland tool room worker:

'As for the tool worker, that's a bit dodgy, saying he's only earning £45 a week, knowing tool workers, the amount of money they get in overtime . . .'

And their decoding of this item is informed by, and leads into a generalised exposition of, a stereotype of the 'greedy car worker/mindless union militants' presumably derived, at least in part, from the media:

'They have got too much power . . . the car unions have got too much power . . . they go on strike, the government puts more money into it [British Leyland], the taxes go up, it doesn't go nowhere . . . They kick up a fuss, the unions, just to sort of show the workers that they're there, they're trying to help them. But if I had a choice I wouldn't join a union . . . to me they're more trouble than they're worth . . .'
'The car workers, they're very petty, they sacked one bloke for nicking, then the whole place comes out on strike, so they've got to re-employ the bloke, pay him all his back pay even though he's been nicking . . .'
' . . . they're one of the worst of the line, the toolroom workers. It's usually them that cause the trouble, them or the blooming paintshop . . .'

Moreover, this stereotype is strengthened in its credibility for them because they can ground it, at least in part, in their own experience; it does have a partial basis in reality (cf. Mepham, 1974; Perkins, 1979).

Q: 'What do you think of the Ford workers?'
'Friend of mine . . . he works at Langley . . . you know. Normally he goes there and has a game of football, this is true, he has a game of football in the machine rooms, muck about, play darts . . . and they're covered for weeks on end . . .'
'They don't work for weeks, because the place is so big, I mean, I know a lot about the Ford workers because I know quite a few people who work there, and they can be covered for weeks on end . . . It's because they're overmanned . . . you can have someone sitting there playing cards all day, it's either your turn or my turn to play cards . . . you could do a bit of mini-cabbing . . . even at my place we've got taxi drivers . . . they're earning a bomb . . .'

This is a view of unions which, whatever its basis in their experience, is resolved at the level of ideology into an acceptance of a dominant stereotype of unions and shop stewards, even at a comic level:

'Did you see that film with Peter Sellers in it? [*I'm All Right, Jack*] The one about unions? . . . really typical, it was on about two Sundays ago . . . absolutely typical, I mean, it's unbelievable – he was the shop steward, sitting on his backside in his little wooden hut all day long, "brothers this and brothers that." '

– clearly an interpellation this group would refuse.

The correlate of this negative disposition towards the unions is a stress on self-reliance and the 'farm worker' section of the programme is decoded in these terms. Rather than sympathy being the dominant frame of reference, here the stress is on the worker's *own* responsibility to 'sort things out' for himself:

'If you think of that farm worker . . . this is my view . . . if you can't afford to live properly . . . wouldn't it be best to have two kids, not three kids? . . . His wife could go out to work . . .'

'. . . you've got to move to a place where there is a lot of work.'

Here this group are employing a perspective to the right of the programme – which most groups see as presenting the farm worker in a quite sympathetic, if patronising, light. When it comes to tax cuts, as proposed for instance by Mr. Worthington, the tax expert, the group's perspective clearly meshes with that proposed by the programme:

'They could reduce all benefits. Unemployed benefit . . . some people *live* on the dole . . . they should pay people below the average living wage so that they're struggling all the time to get a job, that way you [the taxpayer] pay less.'

Here the group share the problematic of the programme at a fundamental level, despite being 'out of position' in relation to the relatively academic mode of address through which Mr. Worthington articulates these propositions:

'I didn't know what he was talking about . . . Well, he goes on, a bit here and a bit there, doesn't give a definite answer.'

However, the decodings of the programme made by this group, despite their cynicism about unions and their stress on work and self-reliance, are not resolvable into a straightforwardly Conservative problematic. Their views exist in contradictory formulations; for, despite the refusal to accept any directly class perspective in relation to unions and 'scroungers' (but see also M. Repo, 1977, on the real basis of working class antagonism to 'social security scroungers'), they also reject and dislike what they see as the programme's partiality towards those who are already 'well off'. In this respect they reject the uncritical presentation of the personnel manager's 'plight':

'They seem particularly concerned with the one that's earning the most money . . . it seemed to be more than just what he was earning, something about his mortgage, and how much he paid out . . .'
'He can always get perks . . . he's got a car on the firm . . . petrol . . . travel expenses . . . it's perks, loopholes. He gets everything . . .'

Moreover, their criticisms of 'the unions' are also articulated not simply as a critique of unionisation in general, but rather as a critique of what they see as a particular, sectional form of unionisation:

'The amount of money there is in the printing at the moment, it's just stupid . . . The unions, they look after Fleet Street, but they're not bothered about anybody else that much . . . They look after Fleet Street . . . more treatment for Fleet Street than anybody else . . .'

Further, what they go on to object to is the consequences of the operation of free collective bargaining for those sections of the workforce who, through no fault of their own, do not have the power of the print or car workers unions:

'People who seem very low paid seem to be the ones doing the most important jobs . . . policemen, firemen . . . ambulance drivers, they do a good job . . . you've got to have those central people . . . got to have the dustmen, otherwise you won't get the rubbish cleared away . . .'

– evidently a perspective which could articulate as well, if not better, with a socialist rather than Conservative political outlook.

Indeed, at the level of 'politics' their views are distinctly equivocal. They clearly do not see themselves as situated within either of the dominant political parties' perspectives. 'A change of government' is seen as a good thing in itself: perhaps a registration of alienation from, rather than confusion within, the field of established party politics:

'It's good to have a change of government anyway . . .'
'Personally I would prefer it if Labour had stayed in, because I mean, I thought they were slowly clearing the trouble up. No government seems to be in long enough to actually clear things up . . .'
'. . . you get another bunch of idiots in and they turn it all upside down, you get another three-day week . . .'
'. . . I don't know how anybody can vote for the Liberals either.'
'. . . I think it's good to have a change of government anyway . . .'
'I think the Labour government will be back in five years.'
'I reckon they will as well, 'cos Thatcher's going to make a right balls-up . . . Well, I think it's good to have a change now and again. 'Cos it's got to the stage now . . . I voted Conservative, I just didn't know what to vote, you know . . . so I voted Conservative, for a change.'

Similarly, their perspective on 'management' is contradictory rather than monolithic:

'Some of the managers are like that . . . worked their way up . . . some go out on all these so-called "business trips" round the golf-course, "business talk" and all this rubbish . . . but some, if they got rid of them, the place would come to a standstill.'

– clearly a different perspective from the militantly anti-management stance of the shop stewards (group 23) as much from the trainee managers' (group 26) monolithic view of management as 'the people who provide this country with everything this country has'.

The equivocation, finally, also applies to the unions: when pressed, their apparently anti-union stance becomes more complex. The point shifts from the attack on the unions *per se* towards a criticism of a particular form of unionisation, a contradictory prescription for 'unions without power':

'Yes . . . it's a good thing to have unions, but not with power . . .'
'You got to have union power, but not too much . . .'
'I mean, when you had small unions, I mean it's better, but you have a big union, it gets worse and worse . . . corrupt, I reckon.'

Group 28

An all male, all black (mainly Nigerian) group of management trainees, aged 22 – 39, mostly with a middle/upper middle class background; predominantly 'don't knows' politically.

This group have little familiarity with British politics or with the codes and cultural conventions of British television, hence discussion moves to an abstract level at which 'taxation' is dismissed as an abstract issue, to be argued over at the level of philosophical principle, as a perennial problem which 'a nation' must face. There is little detailed discussion of the actual programme material shown, because, presumably, much of it is difficult to grasp at the level of connotation – simply in the sense that they do not share, or understand, the framework of taken-for-granted cultural references which underpins the discourse of *Nationwide*. To a large extent, it could be argued that the programme is literally 'unintelligible' to them, beyond the level of the facts and figures given. As 'members' of the English language community they can grasp these aspects of the programme, but as they are not 'members' of the cultural sign community whom *Nationwide* is addressing, they make contradictory and unpredictable decodings of the cultural/connotative messages of the programme.

The dominant response is that:

'The programme is good.'

Indeed they argue that:

'you should watch it because it will make you think more of what is happening in this society . . . if you don't watch such a programme you won't know how people are so affected or what is happening within this society . . .'

Here they perhaps decode the 'place' of *Nationwide* as an 'educational/ serious' current affairs programme, interpret it, rather respectfully, within this context, and accordingly approve it.

They claim to 'agree' with what the programme says (although as we shall see later they hold very contradictory views as to what it did say):

'So I think that what they say in the programme . . . to me it makes a lot of sense . . .'

Moreover, contrary, for instance, to groups such as the trade unionists and the sociology students, who criticise the programme precisely for its unbalanced 'sample' of the population and its exclusion of certain groups, this group, probably because of their lack of experiential criteria with which to judge the issue, applaud its range of representation:

'I enjoyed it because it showed a cross-section of the entire populace.'

The same speaker continues, claiming to 'agree' with the programme, to advance a semi-Marxist theory of taxation which is quite at odds with the interpretation preferred by the programme:

'Yes . . . strictly speaking, taxes should be cut according to needs. From every man according to his ability and to him according to his needs.'

Because of the group's difficulty in placing the cultural messages of the programme, questions on specific items in the programme are reinterpreted as questions about the 'issues in principle', and at the resulting level of abstraction any commitment to 'one side or the other' is evaded. The emphasis is rather on the need to 'think broadly' and 'sit down and plan':

'What I'm getting at . . . cutting taxes wouldn't solve the problem. The problem is you need taxes for hospitals and schools and so on. And you have to give incentives to people who are going to work. There are two sides to the affair and it's a very difficult decision: the government has to sit down and plan.'

'. . . it's a very difficult question, actually . . .'

'What I'm saying now is that the society in general, we are sick because we are not, as I was saying, thinking broadly on things . . . not that I'm on the side of the government or on the side of any industry . . .'

When pressed on the specific question of the programme's presentation of the role of trade unions, the response is again contradictory, starting by situating itself within the preferred mode of the programme:

'. . . I know I'm just new to this country . . . I will say the unions are given too much power, because if the union helped with the job, the government would definitely do something to help the ordinary working man . . .'

But the same speaker continues, without apparently remarking on the contradiction with the preferred reading now proposed:

'. . . the union are the people who are fighting for the right of the ordinary fellow . . . the people who are very close to the citizens are the people who are their representatives. That is why I wouldn't say the unions are being given much power, more than necessary – they're actually doing their job . . .'

This level of contradiction within particular decodings is also present as between the different political perspectives which different members of the group propose. It is rather as if, because their decoding of the programme is so mediated by cultural unfamiliarities, it can coexist with radically different political perspectives: on the one hand a straightforwardly Conservative interpretation of the country's problems:

'If you just go on taxing . . . even if you've got £2m and you take away £1m, then where is the investment? There will be money required for investment, where will it come from, and if you cut out your investment, where are the taxes to come from? . . . A lot of investors have left this country because of taxation . . .'

And on the other hand it can be articulated with a quite contrary Labour perspective, which, interestingly is the one point where the articulation does begin to provide a consistent grip on the specific material of the programme:

'On the programme we had the personnel manager, I think he had everything a man would like to have, a good job, a good home, and telephone and a son at university . . . and yet he wasn't satisfied . . . I mean that's a lot of cash to me, and yet he's complaining . . .'
'. . . to me, I think the top management, they don't need any incentive, because they don't produce anything. It's the skilled workers . . . they are the people that they should give some incentive – because for a man earning £400 p.w., he doesn't need any incentive to do his job because he's producing nothing . . . the man working in Leyland, he said, well they are importing foreign cars into the country. I know what he was talking about, he was talking about losing his job. He said he's been in the job for about 20 years, so if people like him are given incentive, or the farm worker, then definitely he'll do a better job, then the more crops he'll produce at the end of the farming year.'

Group 29

Owing to a fault in the tape recording this group had later to be omitted from the analysis.

6 Comparative Analysis of Differential Group Decodings

Notes on the Pattern of Group Readings

In summary, then (and at the cost of ignoring for the moment contradictory positions within the same groups), the apprentice groups, the schoolboys and the various managers are the groups who most closely inhabit the dominant code of the programme.

The teacher training college students inhabit the 'dominant' end of the spectrum of 'negotiated' readings, with the photography students (inflected by an ideology of 'media' professionalism) and the university students (in a 'Leavisite' version) positioned closer to the 'oppositional' end of the spectrum.

The black students make hardly any connection with the discourse of *Nationwide*. The concerns of *Nationwide* are not the concerns of their world. They do not so much produce an oppositional reading as refuse to read it at all.

The groups involved with the discourse and practice of Trade Unionism produce differentially inflected versions of negotiated and oppositional decodings, depending on their social positions and positioning in educational and political discourses, in combination with those of Trade Unionism.

The problematic proposed here does not attempt to derive decodings directly from social class position or reduce them to it; it is always a question of how social position *plus* particular discourse positions produce specific readings; readings which are structured because the structure of access to different discourses is determined by social position.

At the most obvious level, if we relate decodings to political affiliations then it does appear that the groups dominated by Conservatism – the apprentices, teacher training students and bank managers – produce dominant readings, while those dominated by Labour or socialist discourses are more likely to produce negotiated or oppositional readings. This is not to suggest that it is an undifferentiated 'dominant ideology' which is reproduced and simply accepted or rejected. Rather, it is a question of a specific formulation of that ideology which is articulated through a particular programme discourse and mode of address. Readings are always differentiated into different formulations of dominant and oppositional ideology, and in their differential focus on the ideological problematic and/or the mode of address and discourse of a programme.

The concept of code is always to some extent indefinite. To take the example of dominant code, as employed here it exists in three different versions: for the managers in 'traditional' and 'radical' Conservative forms, for some of the teacher training students in a Leavisite form, and for the apprentice groups in a populist form.

Of these three, according to our characterisation of *Nationwide's* specific ideological problematic, the groups with the closest match of codes to that of the programme are the apprentice groups – that section of the audience that inhabits a populist 'damn-all-politicians' ideology comparable to that of the programme; indeed, decodings are most 'in line' for these groups. For the bank managers and teacher training students the *Nationwide* formulation of dominant code and ideological problematic fails to 'match' their own formulations exactly (for both, from their different perspectives, *Nationwide's* presentation is 'insulting' and therefore to some extent unacceptable). Moreover, there is a disjunction here at the level of 'mode of address': these groups find the *Nationwide* mode unacceptable in terms of a more 'classical' conception of 'good'/serious/educational TV discourse which *Nationwide* does not inhabit.

However, there are problems even with the apprentice groups. Some of these groups (e.g. Group 6) identified with *ATV Today* rather than with *Nationwide,* because of a greater degree of 'fit' with the discourse of *ATV Today* (more human, 'have a good laugh', etc.): an evaluation/identification informed by a populist sense of ITV as more 'entertainment' versus BBC as too 'serious'/middle class.

Conversely, we must distinguish between different forms and formulations of negotiated and oppositional readings, between the 'critique of silence' offered by the black groups, the critical reading (from an 'educational' point of view) elaborated in articulate form by some of the higher education groups, and the different forms of 'politicised' negotiated and oppositional readings made by the trade union groups. (We can also usefully refer here to Richard Dyer's (1977) comments on the problems of dominant and oppositional readings of, for instance, films with a liberal or 'progressive' slant. What Dyer's comments bring in to focus is the extent to which the tripartite 'meaning-systems'/decodings model derived from Parkin assumes that it is dealing always with messages or texts cast within a dominant or reactionary ideological perspective. Dyer's comments on 'aesthetic' and 'substantive' readings are also relevant to the discussion below of the concepts of ideological problematic and mode of address.)

The diagram is presented in this spatial rather than linear form (as in a one-dimensional continuum from oppositional to dominant readings) because the readings cannot be conceived of as being placed along *one* such continuum. For instance, the black F.E. students are not more 'oppositional' than the University students on the same dimension, rather they are operating along a different dimension in their relations to the programme.

The overall 'spread' of the groups' decoding strategies in relation to the two phases of the project is displayed schematically below:

Phase 1: *Nationwide* 19/5/76

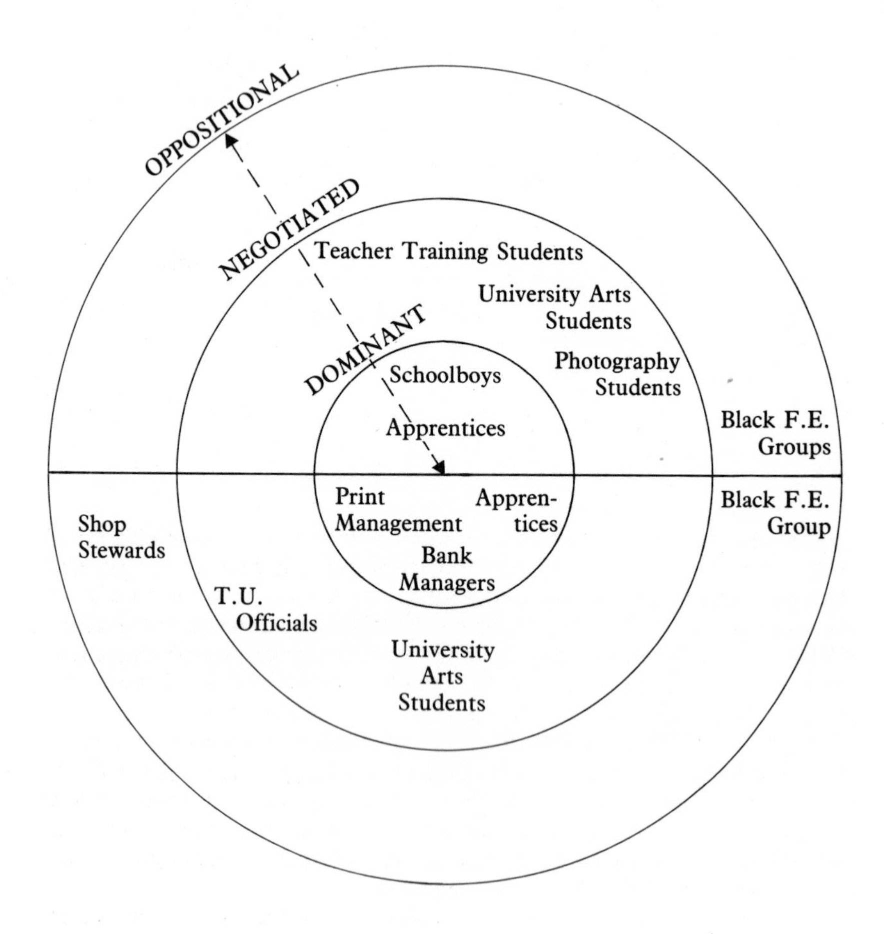

Phase 2: *Nationwide Budget* Special 23/3/77

Thus, in the case of each of the major categories of decoding (dominant, negotiated, oppositional) we can discern different varieties and inflections of what, for purposes of gross comparison only, is termed the same 'code':

Forms of Dominant Code

Groups	
26	Print Management: radical Conservative
21, 24	Bank Management: traditional Conservative
1-6 & 27	Apprentices: Populist-Conservative/cynical
10, 12	Schoolboys: deferential (?)

Forms of Negotiated Code

14, 15	Teacher Training College Students: Conservative Leavisite
7, 19	University Arts Students: radical Leavisite
8, 18	Photography Students: technicist 'professional' perspective
20, 22	Trade Union Officials: Labourist 'official' perspective

Forms of Oppositional Code

11, 13, 16, 17, 25	Black F.E. Students: alienated 'critique of silence': sub-cultural perspective
23	Shop Stewards: radical rank & file perspective: class perspective

Thus, social position in no way directly correlates with decodings – the apprentice groups, the trade union/shop stewards groups and the black F.E. student groups all share a common class position, but their decodings are inflected in different directions by the influence of the discourses and institutions in which they are situated. In one case a tradition of mainstream working class populism, in another that of trade union and Labour party politics, in another the influence of black youth cultures.

Indeed, in the case of the decodings of the 19/5/76 programme, with the exception of the black groups (who arrive at an oppositional perspective through a different, sub-cultural route) the education system inflects the groups' decodings in a pattern such that it is those more highly placed in that system (and in general those with a more middle class background) who come closer to an oppositional perspective.

Interestingly, in the case of the decodings of the 'Budget' programme (which deals much more directly with issues of class and politics) with the exception of the apprentice group, whose insertion in the mainstream/populist tradition inflects their reading towards the dominant perspective, it is here that we find a greater convergence of middle class positions with dominant or negotiated perspectives and working class positions with more oppositional readings. This is true even within one category: the university arts group in Phase 1 produces negotiated readings with a distinctly 'oppositional' tinge, while the similar group in Phase 2 produces a negotiated reading much closer to the 'dominant' edge of that category, once it is confronted directly with issues of a specifically 'politico-economic' rather than 'cultural' relevance.

Moreover, there are of course some differences between groups from the same category – between the different apprentice groups, or between the different trade union groups, for instance; differences which would require for their systematic investigation much more in the way of time and resources than were available. I can only reiterate my view that these are subsidiary differences, within the perspective taken, as are the differences between individual readings within each group – differences which are to be acknowledged, but which, I would argue, do not erase the patterns of consistency and similarity of perspectives within groups which I have attempted to establish at a more fundamental level.

In order to attempt to demonstrate this claim I now provide, by way of overview, some recapitulation of the basic levels of differential decodings between groups in the different categories.

Inter-Group Differences: An Overview

1 Apprentice Groups

These working class groups inhabit a discourse dominated on the one hand by Conservatism and on the other by a populism which rejects the whole system of party politics (they are militant non-voters: 'they're all as bad as each other/they've all got the same ideas'), and to some extent identify with the National Front. (Note: these groups are the only ones in my sample which showed any NF influence.) The tone of their overall response to the programme is one of cynicism and alienation.

However, despite the overall tone of rejection and cynicism, most of the main items in the programme are decoded by these groups *within* the dominant framework or preferred reading established by the programme. The situation here seems to be the converse of that outlined by Parkin in his account of working class groups who accept a dominant ideology at an abstract or generalised level but reject it, or 'negotiate' its application, in their own specific situation. Here, on the contrary, we have working class groups who proclaim a cynical distanciation from the programme at a general level but who accept and reproduce its ideological formulations of specific issues.

This disjunction may be conceptualised in two different ways. In the first place, we might hypothesise that the general tenor of cynical response amounts to a defensive stance. It is an attempt to appear worldly-wise, not to be taken in by television – a stance which, because it is not produced by adherence to any definite, ideological problematic other than that provided by the programme, fails to deliver any alternative formulations in which the particular issues dealt with can be articulated. It thus returns the viewer, by default, to acceptance of *Nationwide's* formulations. This, then, is a peculiarly weightless cynicism which combines quite readily with an acceptance of the framework of meanings established in the programme.

There is a second way of conceptualising the disjunction noted above. We can adapt Steve Neale's (1977) distinction between the ideological problematic of a text – 'the field and range of its representational possibilities' and the mode of address of a text – its relation to, and positioning of, its audience. The concept 'ideological problematic' designates not a set of 'contents' but rather a defined space of operation, the way a problematic selects from, conceives and organises its field of reference. This then constitutes a particular agenda of issues which are visible or invisible, or a repertoire of questions which are asked or not asked. The problematic is importantly defined in the negative – as those questions or issues which cannot (easily) be put within a particular problematic – and in the positive as that set of questions or issues which constitute the dominant or preferred 'themes' of a programme.

The concept of 'mode of address' designates the specific communicative forms and practices of a programme which constitute what would be referred to in literary criticism as its 'tone' or 'style'. In Voloshinov's terms these are the 'occasional forms of the utterance', those forms which are deemed situationally appropriate. This 'appropriateness' is principally defined in relation to the programme's conception of its audience. The mode of address establishes the form of the relation which the programme proposes to/with its audience. Thus the disjunction noted earlier must also be characterised as one where these groups reject at a general level the programme's mode of address or articulation – as too formal/middle-class/ 'BBC-traditional' – but still inhabit the same 'populist' ideological problematic of the programme, and thus decode specific items in line with the preferred reading encoded in the text.

The sharing of this ideological problematic is particularly difficult to investigate or articulate precisely because it is the problematic of 'common sense' and as such is deemed obvious and natural. However, the point is that 'common sense' always has a particular historical formulation; it is always a particular combination constituted out of elements from various ideological fields and discourses – what is shared here is a particular definition of 'common sense'.

These groups are at times hostile to the questions being asked, annoyed at being asked about what is obvious; it seems hard for them to articulate things which are so obvious. There is also a defensive strategy at work. Judgement words such as 'better', 'boring', etc. are used without explication, and explication is impossible or refused because 'it's only common sense, isn't it?' *Nationwide's* questions are justified as 'natural', 'obvious' – and therefore unproblematic: 'They just said the obvious comment, didn't they?'

The *Nationwide* team are seen as 'just doing a job', a job seen in technical terms, dealing with technical communication problems. To ask questions about the socio-political effects of *Nationwide's* practices is seen as going 'a bit too far in that . . . approach . . . it's going a bit too deep really . . . after all people don't sit by the fireside every night discussing programmes like we're doing now, do they?'

In these discussions, the terms of criticism are themselves often derived from the broadcasters' own professional ideology. Thus the most critical term is precisely that of bias or balance: 'They're biased though, aren't they . . . ' 'it biased it before it even started.' But the *limit* of criticism is also set by the problematic of 'bias': specific items may be biased, at exceptional moments 'they' might 'slip in the odd comment, change it a bit', but these are merely exceptions in a balanced world, produced by programme-makers just doing a job like everyone else. This is the ideological structure described by Poulantzas (1973) as one whereby:

> 'The dominated classes live their conditions of political existence through the forms of the dominant political discourses: this means they often live even their revolt against the domination of the system within the frame of reference of the dominant legitimacy.' (p. 223)

When I began the research I expected to find a clear division so that decoding practices would either be unconscious (not recognising the mechanisms of construction of preferred readings) and as such, in line with the dominant code or else, if conscious, they would recognise the construction of preferred readings and reject them. In fact, the recognition of 'preferring' mechanisms is widespread in the groups* and combines with either acceptance or rejection of the encoded preferred reading; the awareness of the construction by no means entails the rejection of what is constructed.

This is to point to the distinction (cf. Stevens, 1978, pp. 18-19) between the level of signification at which the subject is positioned in the signifying chain; and the level of the ideological at which particular problematics are reproduced and accepted or rejected. The two levels can actually operate in variable combinations: on the one hand, the positioning of the subject does not guarantee the reproduction of ideology (the same subject position produced from/by the text can sustain different ideological problematics); conversely, a break in, or deconstruction of, the signifying mechanisms of the programme (an interview 'seen through' as 'loading the questions', etc.) will not automatically produce or necessitate an oppositional reading, if there remains an underlying shared and accepted ideological problematic.

*This in fact provides an interesting extension of John O. Thompson's (1978) comments in his review of *Everyday Television: 'Nationwide'*, where he questions the extent to which this project has itself failed at times to escape the trap whereby only the media analyst is assumed, for some unspecified reason, to be beyond the influence of the 'dominant ideology'.

2 Trade Union Groups

Contrary to Blumler and Ewbank's study, which found little difference in decoding between ordinary trade union members and full time officials, my material shows the considerable influence on decoding of different kinds and degrees of involvement in the discourse and practices of trade unionism.

Comparing groups 1–6 and 20, 22, 23 – groups with the same basic working class, socio-economic and educational background – there is a profound difference between the groups who are non-union, or are simply *members* of trade unions, and those with an active involvement in and commitment to trade unionism, the latter (groups 20, 22, 23) producing much more negotiated or oppositional readings of *Nationwide*. So the structure of decoding is not a simple result of class position, but rather (cf. Willemen, op. cit.) the result of differential involvement and positioning in discourse formations.

However, this is not a simple and undifferentiated effect. There are significant differences between the three trade union groups as well, which need to be explored in terms of the differential involvement in discourses which produce these readings. There are, on the one hand, differences between the two groups of full-time trade union officials – group 22 producing a noticeably more 'oppositional' reading than group 20 – which we can only, hypothetically, ascribe to the effects of working in different unions, differentially placed in the public/private and productive/service sectors – in short to differences within the discourse of trade union politics.

Further, there are the significant differences between the articulate, fully oppositional, readings produced by the shop stewards as compared with the negotiated/oppositional readings produced by groups 20 and 22 taken together (full-time officials). This, I would suggest, is to be accounted for by the extent to which, as stewards, group 23 are not subject so directly to the pressures of incorporation focused on full time officials and thus tend to inhabit a more 'left-wing' interpretation of trade unionism. The difference between these groups also has to be accounted for by the effect for group 23 of their on-going involvement in a highly politicised educational situation, as part-time students on a labour studies diploma run by the student collective. This educational context in combination with the experience of working as a shop steward is probably responsible for the development of such a coherent and articulate oppositional reading of the programme as this group makes.

3 Teacher Training College Groups

While the teacher training students (groups 14 and 15) share with the apprentices (1–6) a dominant political affiliation to the Conservative party (although already differently inflected in each case), their insertion in the discourse of Higher Education acts, in combination with this 'Conservatism', to shift their readings further into the 'negotiated' as opposed to 'dominant' area. It is this discourse which differentiates their situation, and their readings, from those of the other 'conservative' groups.

Taking the involvement in educational discourse as a variable we can compare the decodings (manifested for instance in differential use of the term 'detail' as a value-judgement/criterion by which programmes are assessed) of groups 14 and 15 – students training to be teachers, heavily implicated in the discourse of education – with groups 11, 13, 16 – black students resistant or without access to this particular discourse, with its high estimation of serious, educational TV and its concern for information and detail. From the black students' perspective the programme is seen to go 'right down into detail' and as a consequence is 'boring' and fails to live up to their criterion of 'good TV' as principally entertaining and enjoyable. For the teacher training students the *Nationwide* programme fails because it does not have enough detail or information and fails to be serious or worthwhile.

The differentially evaluative use of the same term 'detail' noted above, as it functions in different discourses, is a perfect example of Voloshinov's concept of the 'multi-accentuality' of the sign, and the struggle within and between different discourses to incorporate the sign into different discursive formations.

The differential involvement in the discourse of formal education of the teacher training student groups and the black 'non-academic' students groups can precisely account for their responses to the programme and for the different framework within which they articulate and justify these responses. Moreover, the notion of differential involvement in educational discourse can be mobilised to explain the decoding of the students 'rubbish' item by some of the teacher training students groups as opposed to most of the other groups. The groups who inhabit this same eductional discourse tend to read sense into the students item; the other groups tend to accept *Nationwide's* implicit characterisation of the students as 'wasting time'.

The comparison of perspectives is at its sharpest in the case of these groups because they stand at opposite ends of the spectrum of involvement in educational discourse. Trainee teachers groups 14 and 15 are probably those most committed to that discourse in the whole sample, as the working class black groups (11, 13, 16) are probably those most alienated from the discourse of formal education.

4 Black Further Education Students

These groups are so totally alienated from the discourse of *Nationwide* that their response is in the first instance 'a critique of silence', rather than an oppositional reading; indeed, in so far as they make *any* sense at all of the items, some of them at times come close to accepting the programme's own definitions. For example on Meehan: 'all I heard was that he just come out of prison . . . something he didn't do, that's all I heard.' In a sense they fail, or refuse, to engage with the discourse of the programme enough to deconstruct or redefine it. However, their response to particular items is, in fact, overdetermined by their alienation from the programme as a whole.

Taking up the argument (from Carswell and Rommetveit, 1971, 'Social Contexts of Messages') that the action of any one discourse and its interpretation always involves the action of other discourses in which the subject (as the site of interdiscourse) is involved, the differential interpretation (and rejection) of *Nationwide* by the predominantly black groups (11, 13, 16) is not to be explained as the result of some interruption of the 'natural flow' of communication (cf. Hall's, 1973, criticisms of this model of 'kinks' in the communication circuit). Rather, what this points to is that these groups, not simply by virtue of being black, but by virtue of their insertion in particular cultural/discursive forms, and their rejection of or exclusion from others, (which insertion is structured, although not mechanistically determined, within their biographies, by the fact of being born black) are unable or unwilling to produce 'representations' which correspond to those of the programme, or to make identifications with any of the positions or persons offered through the programme.

As Henry puts it ('On processing of message referents in Contexts', in Carswell and Rommetveit, op. cit.):

> '. . . Differences between representations indicate differences in the analysis of the real, assumed or imagined events, situations . . . which the message is about, as well as differences in the positions of speakers. They can also indicate differences in the loci occupied by different speakers within the social structure and hence also differences in their economic, political and ideological positions . . .
> . . . in order to intrepret the messages he receives, the addressee must elaborate representations with those messages . . . if the addressee is not able to build up such representations the message is meaningless for him.' (pp. 90–91)

Here we have a clear case of disjunction between the 'representations' of the programme and those of these black students' culture. However, this 'effect' – which appears as 'disruption' of communication – must alert us to a wider question. Here the action of the cultures and discourses which these groups are involved in acts to block or inflect their interpellation by the discourse of *Nationwide*; as a negative factor this is quite visible. We must note, however, the converse of this argument in the cases of those groups (all the others except 11, 13, 16) who are not involved in these particular discourses and therefore do not experience this particular blocking or inflection of *Nationwide* discourse. Here it is not simply a case of the absence of 'contradictory' discourses; rather it is the presence of other discourses which work in parallel with those of the programme – enabling these groups to produce 'corresponding' representations. Positively or negatively, other discourses are always involved in the relation of text and subject, although their action is more visible when it is a case of negative – contradictory rather than positive – reinforcing effect.

5 Higher Education Students

It is among these groups, inserted in the discourse of higher education but unlike groups 14 and 15 away from the conservatism of the teacher training college, that we find an articulate set of negotiated and oppositional readings and of redefinitions of the framework of interpretation proposed in the programme. Moreover, because of their particular educational backgrounds they consistently produce deconstructed readings; that is to say, they are particularly conscious of the methods through which the *Nationwide* discourse is constructed.

These groups, rather like the bank managers at times, dismiss *Nationwide*'s style and mode of address as simply 'undemanding entertainment/ teatime stuff', again criticising *Nationwide* from a framework of values (in their case a Leavisite notion of 'high culture') identified with those of 'serious BBC current affairs'.

As argued above, their decodings of the 'Budget' programme (as opposed to their decodings of the 'Everyday' programme) are consistently less oppositional. In that programme, in the sections with the tax expert and the three families, they make, again like the managers, little or no comment on the problematic employed by the programme. They criticise the style as 'patronising' and further, also like the managers, see no *particular* class theme in the 'families' section – while accepting *Nationwide*'s reduction of class structure to individual differences: 'different classes . . . it's a fact of life . . .' I would argue that this lack of comment is evidence of the invisibility rather than the absence of themes, an invisibility produced by the equivalence between the group's problematic and that of the programme in this respect.

Code	*Group*	*Mode of Address*	*Ideological Problematic*
Dominant	Bank Managers	Extensive critical* comment: focused	Little comment – invisible/ non-controversial/shared
Negotiated/ Oppositional	Trade Unionists	Treated as subordinate issue	Rejection – focus on the problematic*

* It is interesting to compare these findings with those quoted in Sylvia Harvey (1978), p. 147. Paul Seban (author of the film *La CGT en mai '68*) is discussing the differential responses to screenings of the film on the part of management and workers:

> 'This has often been reported to me. Workers see the film and say: "That's our strike." Engineers and office staff see it and say: "It's well made. The images are beautiful." '

Dominant and Oppositional Readings: Mode of Address and Ideological Problematic

The differential decodings of the bank managers as opposed to the trade unionists, represented diagramatically opposite, raise important theoretical problems concerning the nature of decoding. If we contrast these two groups of readings across the two dimensions of communication defined by Neale (1977) as mode of address and ideological problematic we see that the dominant and oppositional readings focus on quite opposite aspects of the programme.

Interestingly, the managers hardly comment at all on the substance of the ideological problematic embedded in the programme. Their attention focuses almost exclusively on the programme's mode of address, which they reject as 'just a tea-time entertainment programme, embarassing . . . patronising . . . exploiting raw emotion . . . sensationalism'. Their adherence is to a mode of address identifiable as 'serious current affairs'; they mention the *Daily Telegraph, Panorama* and the *Money Programme* as models of 'good coverage' of these issues, and dismiss *Nationwide* in so far as it fails to live up to the criteria established by this framework.

By contrast, the shop stewards can accept the programme's mode of address to some extent: 'It's light entertainment/not too heavy/easy watching/quite good entertainment': what they reject is *Nationwide*'s ideological formulation of the 'issues'. Thus in the case of the 'Budget' programme the dominant readings concentrate their comment (which is largely critical) on the programme's unacceptable style or mode of address, while for them the ideological problematic passes invisibly, non-controversially; whereas the oppositional readings focus immediately on the unacceptable ideological problematic, and the mode of address is treated as a subordinate issue and given little comment – or even appreciated.

To theorise these findings we can extend the argument made by Ian Connell in respect of the transparency of programme discourse to cover the case also of the transparency of an ideological problematic in cases of 'fit' between the problematic employed in the encoding and that of the decoding. As Connell (1978) argues in respect of programme discourse:

> 'We must ask whether these roles or "spaces" in the discourse are accepted, modified or rejected, by whom and for what reasons. We shall take as an indicator of acceptance, for example, the absence of any spontaneous comment about the discourse as such, and about the roles advanced in it. In other words, for those who share the professional sequencing rules the discourse will not "intrude", and they will spontaneously talk about the topic . . . When there is modification or rejection, we would expect explicit comments concerning the organisation of the discourse.'

Similarly, for those who share the ideological problematic of the programme the problematic does not intrude (indeed the bank managers and students deny the very presence of any particular problematic) and they spontaneously talk about the discourse or mode of address. Where groups, such as the trade unionists, do not share the problematic it is quite 'visible' because it is 'controversial' to them. Presumably, in the limiting case of exact 'fit' between encoding and decoding in both mode of address and ideological problematic the whole process would be so transparent/non-controversial as to provoke no spontaneous comment at all.

This is to propose the general principle that the unasserted (what is taken as obvious/natural/common-sensical) precedes and dominates the asserted (particular ideological positions advanced within this taken-for-granted framework). As long as the (unasserted) 'frame' is shared between encoder and decoder then the passage of the problematic embodied in that frame is transparent. We can then speak of four decoding positions, represented diagrammatically below:

1) Where the problematic is unasserted and shared, and passes transparently (e.g. the unstated premise in a report that 'race' is a problem – which premise is 'unconsciously' shared by the decoder).

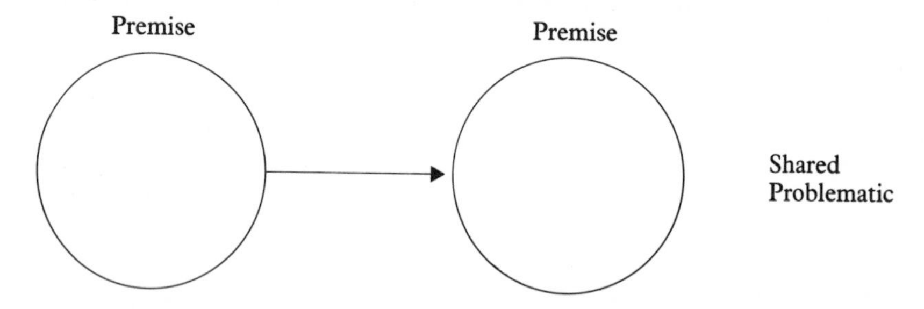

2) Where a particular position within a problematic is asserted and accepted; here the encoded position is accepted by the decoder but it is consciously registered as a position (not a 'natural fact') against other positions. To the extent that this is then a recognition of the necessary partiality of any position it is a weaker structure than 1 (e.g. the explicitly made and accepted statement that blacks cause unemployment).

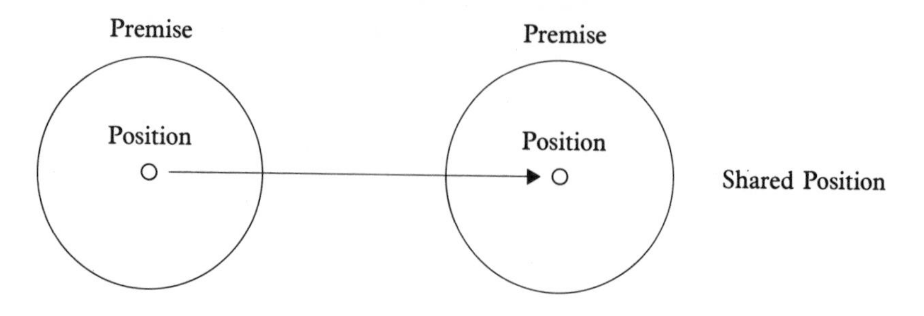

3) Where a particular position within a problematic is asserted but rejected, while the problematic itself is not brought into question (e.g. the explicitly made statement that blacks cause unemployment is rejected as simply another of the politicians' endless excuses for their failures and the racist problematic is not necessarily challenged).

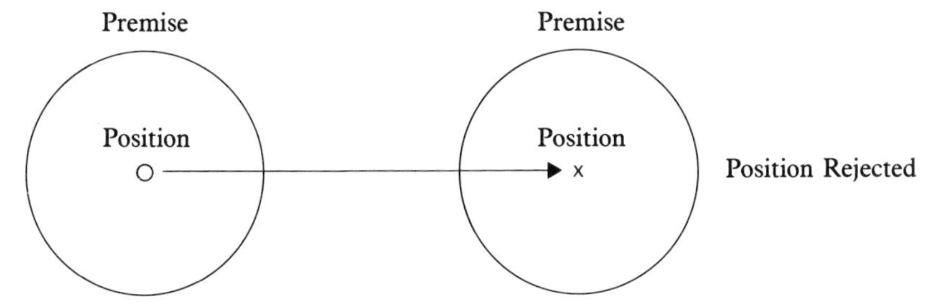

4) Where the underlying problematic is consciously registered and rejected (e.g. a particular report with racist premises is deconstructed to reveal those premises and another problematic is inserted in its place).

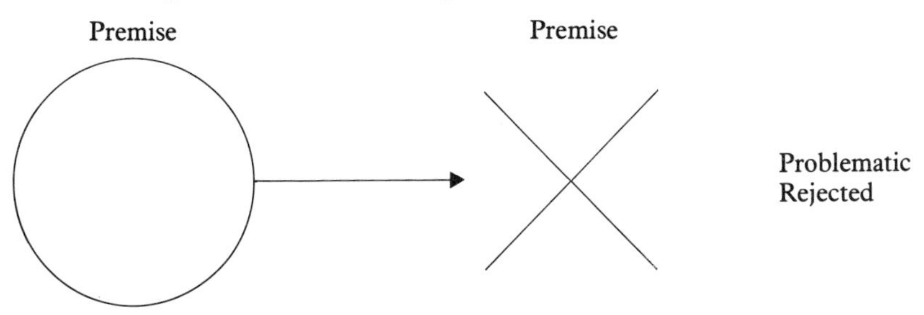

7 Decoding Television: Theorising Determinations

The Effectivity of the Text

It seems that media and film studies are still subject to the kind of oscillation described in Chapter 1. At one moment the field is dominated by a theory (such as 'uses and gratifications' in recent years) which holds the media to have little or no direct 'effect' on its audiences, and at the next moment the pendulum swings towards the dominance of a theory (such as that developed in *Screen* more recently) of the near total effectivity of the text, in their terms, in the 'positioning of the subject'. In order to escape from this oscillation we need to develop a theory which gives due weight to both the 'text' and 'audience' halves of the equation.

Having situated the project in relation to the dominant paradigms of the sociology of mass communications in the earlier chapters, I shall concentrate here on the relation of this project to that body of film theory developed predominantly in *Screen*, and in particular, focusing on the audience ('subject')/text relation proposed there. There is space here to discuss that body of work only in its broad outlines; to this extent I may be faulted for presenting this theory as more monolithic than it in fact is. It is of crucial importance here to note the contributions of Neale (1977) and Willemen (1978), who while writing within the framework of *Screen* seem to raise problems which highlight the deficiencies of the 'orthodoxy' developed by Heath, MacCabe and others and lead in the direction of possible solutions to these problems.

As we have seen in the research outlined at length in the previous chapters, a given text can be appropriated by readers in different ways, read to produce divergent interpretations. Thus the text does not construct or determine the reader – that is to grant complete determinacy to the text and hence to conceive of readers as:

> 'blanks, mere functions supporting the text, and passively to be constructed and put into place according to the whim of the text.' (Willemen, op. cit., p. 49)

This is the point of Steve Neale's argument that textual characteristics are no guarantee of meaning. In the case he examined, texts without the characteristic textual mode of address of propaganda were able to function as propaganda in certain circumstances, and, conversely, texts with a propagandist mode of address could be 'defused' by the context in which they operated.

However, 'interpretations' are not arbitrary but are subject to constraints contained with the text itself. Textual characteristics do have effectivity, though not absolute effectivity; for:

> 'Some texts can be more or less recalcitrant if pulled into a particular field, while others can be fitted comfortably into it.
>
> Texts can restrict readings (offer resistance)- although they can't determine them. They can hinder the productivity of the plurality of discourses at play in them, they can emphasise certain discourses as opposed to others (through repetition or other "foregrounding" devices).' (Willemen, p. 63)

In recent years *Screen* has made a consistent intervention into the field of film and media studies which has usefully focused on a number of important questions concerning the formal operations of ideology. However, I would argue that the problematic proposed there is often flawed precisely by a mechanistic assumption that the audience/viewer is necessarily 'positioned' or placed by a programme so as to reproduce the dominant ideology. Their concern is with the (re)production of the ideological subject by the text, but the problematic is overly deterministic:

> 'an important – determining – part of ideological systems is then the achievement of a number of machines (institutions) that can move the individual as subject.' (Heath, 1976)
> 'In the brief action of finding Vincent, Linda and Keith in this section the viewer is placed in the enounced, in the enunciation and in the setting of the unity of the two, the programme's process – and – position.' (Heath and Skirrow, 1977)
> 'it maps out little fictions . . . in which the viewer as subject is carried along.' (op. cit., p. 58)
> '. . . to expose TV in its programmes . . . has a value against the sufficiency of the "institution and its unquestioned performance of the subject".' (op. cit., p. 59)

I would argue that we need precisely to question that 'performance'. Heath and Skirrow's formulation *assumes* the efficacy of the programme's structure and strategy in placing its audiences into a given position in relation to a text. What in the formulation of 'preferred readings' is presented as a problematic attempt at the structuration of meaning for the audience is here presented as a necessary result. I would not argue with Heath and Skirrow's formulation of TV as a medium where:

> 'the viewer doesn't make his own decision what to look at: it is made for him.' (op. cit., p. 33)

In *Nationwide*, for example, a programme link may take a form such as:

'And after your own programme, we go racing . . .'

– and the point is that 'we' are implicated, because if they go racing, we go racing, unless we switch off. But this is already to grant the space between text and audience. The presenter attempts to implicate us, as audience – but it is an attempt which may be more or less successful; it is not a pre-given and guaranteed achievement.

Here I would want to insist on the active nature of readings and of cultural production. Too often the audience subject is reduced to the status of an automated puppet pulled by the strings of the text. As Connell (1979) correctly argues:

> 'Any reading is never simply consumptive, a mere adoption of the subject-forms mapped out by the formal operations of a text. Any reading is also, and always, interrogative. In short, the reading of a text is, at all times, a practice of "expanded reproduction" (Marx), a reconstruction of the text in and through extra-textual discursive formations.' (p. 133)

At this point I would like to take up some of the implications of Paul Willis' work on forms of working class cultural opposition within the school system, work which gives due weight to the uneveness of those practices which act to reproduce social structures, and their contradictions. As Willis (1978) argues, the problem is that:

> 'Structuralist theories of reproduction present the dominant ideology . . . as impenetrable. Everything fits too neatly. There are no cracks in the billiard ball smoothness of process . . .' (p. 175)

As he goes on to say:

> 'Social agents are not passive bearers of ideology, but active appropriators who reproduce existing structures only through struggle, contestation and partial penetration of those structures . . .' (p. 175)

The point is that dominant ideological and cultural forms are:

> 'not produced by simple outside determination [but] . . . also from the activities and struggles of each new generation . . . which actively produce and reproduce what we think of as aspects of structure. It is only by passing through this moment that determinations are made effective in the social world at all.' (pp. 120–1)

This is also to say that, if we are to speak of the reproduction of a dominant ideology, we must see that such an ideology can *only* have effectivity in articulation with the existing forms of common sense and culture of the groups to whom it is addressed.

However, this is not to argue that subordinate groups are free to produce their own cultural life and forms, or readings of dominant forms, in unlimited space. Parkin effectively makes the point that:

'values are not imposed on men in some mechanistic way. Men also impose their will by selecting . . . from the range of values that any complex society generates. At the same time individuals do not construct their social worlds in terms of a wholly personal vision, and without drawing heavily on the organised concepts which are part of public meaning systems. Variations in the structure of attitudes of groups or individuals . . . are thus to some extent dependent on differences in access to these meaning systems.'

This is the process to which Willis refers when he argues that subordinate groups produce their own cultures 'in a struggle with the constrictions of the available forms' (p. 124) – a process which is not 'carried' passively by *tragers*/agents but contradictorily produced out of a differentially limited repertoire of available culture forms.

Signification and Ideology: the Problem of Specificity

I would argue that Hirst and his followers, in their concern with the specificity of 'signifying practices', have themselves consistently tended to reduce ideology to signification; to reduce, for instance, a particular television text to an example of 'Television-in-General', or 'Televisionality'.

Tony Stevens (1978) has recently criticised the 'Encoding/Decoding' model for assuming 'more of a "lack of fit" between ideology and signification than is necessary to grasp the significance of their non-identity'. This argument seems to me to depend on the unjustified equation of the reproduction of ideology with the production of subject positions. Moreover, this is linked to his second criticism, that the model 'conceives the text in terms of the transmission rather than the production of messages'. But what he seems to want to substitute is a model in which the text is the total site of the production of meaning rather than working within a field of pre-existing social representations, and the production of the subject is placed entirely on the side of signification, ignoring the social construction of the subject outside the text. This connects with the third criticism, where the distinction between the social subject and the reader of the text in the Encoding/Decoding model leads Stevens to argue that the model 'admits no way in which the text constructs its own reader and in this denies the very nature of reading'. On the contrary, the model's conception of the preferred reading of a text relates precisely to this point – but it is not assumed here, as it is usually within the *Screen* problematic, that the preferred reading will necessarily be taken up by the reader of the text.

An alternative way of conceptualising the problem of 'specificity' is to pose the concept of professional code as precisely being the theoretical space where the 'specificity' of TV as a medium is to be situated, and is to be seen to have determinate effects (*pace* Hirst) in terms of a sedimented set of professional practices that define how the medium is to be used (i.e. it is not a quality of the medium as such). The point here, however, is that this 'specificity' is not a world apart from that of ideology but is always articulated with and through it. This articulation is precisely the space in which we can speak of the relative autonomy of the media.

An example from the programme used in Phase 1 of the project may help to illustrate the point. The Meehan item is opaque – hard to read. What *Nationwide* have chosen is a particular focus on the 'human', subjective or expressive aspect of that person's experience. What is 'not relevant' as far as they are concerned is the whole political background to the case. Now that is not to say that this is a straightforwardly ideological decision to block out the political implications of the case. It's much more, in their terms, a communicative decision, as it appears to them; that is their notion of 'good television', to deal in that kind of 'personal drama'. This is an example of the operation of 'professional codes': a case where a dominant code or ideology has to pass through the concerns of a professional code. The decision to focus on a close-up of Meehan's face is a decision within the terms of a professional code. Its consequence is that the 'sense' of the case, of the background to the item, is virtually impossible to 'read' and the viewer is made more than usually dependent for clarification on the presenter's 'summing up' of the item: the 'professional' decision thus has ideological consequences.

Within the terrain mapped out by *Screen*, the concern with analysing the specificity of televisual signifying practice often gets re-written as a concern:

> 'to discover television itself.' (Caughie, 1977, p. 76)
> 'to insert the question of television itself.' (Heath and Skirrow, 1977, p. 9)

The concept of 'television itself' is on a par with the concept of 'ideology in general' or 'production in general': as an analytic abstraction which cannot be related directly, without further socio-historical specification, to the business of analysing the production and reproduction of specific ideologies in concrete social formations at a specific stage in their historical development.

In fact Heath and Skirrow themselves state that:

> 'it is only in the detailed consideration of particular instances that the effective reality of television production can be grasped.' (p. 9)

but this seems to sit unhappily with the formulations quoted above, which indicate a tendency to abstraction or hypostatisation – no doubt informed by a concern with 'specificity' – of the level of signifying practice, to the point where it becomes unclear what relations these practices contract with other levels of the social formation.

Heath and Skirrow seem to go further. They argue that 'the level of social connotations' is not their concern; here the concern with 'television itself' leads to a position where:

'what counts – initially – institutionally, is the communicationality [not] the specific ideological position.'

Indeed they argue that it is of:

'little matter in this respect what is communicated, the crux is the creation and maintenance of the communicating situation and the realisation of the viewer as subject in that situation.' (p. 56)

The question of 'what is communicated' is, however, not so simply to be relegated to a subordinate position. *Nationwide*, for example, is in the business of articulating 'topical' ideological themes for its audience, and if we are to understand the specificity of *Nationwide* discourse, we need to pay attention to 'what is communicated' within the general framework of the 'reproduction of the communicative situation and the realisation of the viewer as subject in that situation'. (Heath and Skirrow, p. 56)

As ideologies arise in and mediate social practices, the 'texts' produced by television must be read in their social existence, as televisual texts but also as televisual texts drawing on 'existing social representations' within a field of dominant and preferred ideological meanings. For this reason a concern with the 'level of social connotations' (Heath and Skirrow, p. 13) must remain as a central part of the analysis. It is only through an analysis of the historical specificity of ideological themes, as articulated by particular programmes in specific periods, that we can begin to map (albeit perhaps only descriptively) the shifting field of hegemony and ideological struggle which is the terrain in and on which the specific practices of television are exercised. This is to argue for an articulation of discursive/signifying practice with political/ideological relations much closer than that proposed in the problematic of a concern with 'television itself'.

Alan O'Shea (1979) has usefully introduced Voloshinov's (1973) distinction between 'themes' and 'mechanisms' (presented in *Marxism and the Philosophy of Language*, p. 100) to deal with the articulation between the question of what an ideology is and how it comes to be appropriated. As he rightly argues:

'themes are only articulated through the mechanisms – which have a determining effect on what the themes are – but . . . the study of the mechanisms . . . only bears fruit when analysed in conjunction with what is represented.'

The *Screen* problematic focuses exclusively on the process of signification as the production of the subject. But signification always occurs in ideology, and when the passage of dominant meanings is successful this is so as a result of operations at both levels. Not simply because of the successful positioning of the subject in the signifying process (the same signification or position is compatible with different ideological problematics; successful positioning in the chain of signification is no guarantee of dominant decodings); but also because of the acceptance of *what* is said.

The 'What' of Ideology

This is to insist on the importance of 'what gets said' and also on the question of 'what identifications get made by whom'. MacCabe (1976) attempts to introduce into the *Screen* problematic the question of 'readers' who are offered 'explanations' of their lives, ideological viewpoints which they may or may not adopt. The difficulty is that the established *Screen* problematic provides no theoretical space into which this question can be inserted – as Adam Mills and Andy Lowe (1978) point out in their critique of MacCabe:

> 'If ideology is posed as a structuring moment then the matter of content is irrelevant, since the ideological position remains the same (regardless of content) . . . if on the other hand MacCabe proposed that the ideological position is given not by . . . identification as process, but by what *specific* imaginary is constituted (i.e. regardless of process) then this must have, necessarily, radical effects for what is now proposed as the body of concepts relevant for this work. . . . up until now the subject . . . has been an effect of the signifying chain, as a mechanism, as a regulated process . . . But ideology is now presented as an explanation, which must deal with contents, and to which an individual (not apparently as subject) can maintain a critical distance and exercise a choice. The vital question is, by what means? In all the preceding analyses there has been no provision for an individual entering into any relation with signification as anything but a subject. The individual can be present in discourse only as a subject, and the ordering of that discourse has permitted no contradiction in determining the form of that subjectivity.'

In order to provide the theoretical space in which these concerns can be inserted it is necessary to amend the terms of the problematic. It has to be accepted that the matter of 'what specific imaginary is communicated' is important both positively and negatively – in cases of both successful and unsuccessful interpellation – not simply as a negative consideration that at times 'interrupts' the process of signification.

Jim Grealy (1977) puts this well in relation to the ideological effectivity of Hollywood cinema. He argues that popular cinema cannot:

> 'avoid the real concerns of [its] audience. Hollywood cinema elaborates a certain understanding of the common experiences of the individual subject in capitalist, patriarchal society and it is this ability to elaborate conflicts which have a real basis in the daily struggles of the individual subject/spectator, and its ability to propose representations/fantasies in which the spectator can find imaginary unity that allows Hollywood to find a popular basis and to carry an important ideological charge.' (p. 9)

To return the argument to the Heath and Skirrow piece on *World in Action*, Alan O'Shea contends that the authors, in their exclusive concentration on the form of signification, leave:

> 'no place for the other "circuit", the viewer as bearer of a complex of interpellations, many of which are not constructed by the TV text.'

To take an example from the *World in Action* analysis, the authors argue that Dudley Fiske – the 'expert' called in to sum up on the problem of truancy – has the power that he does within the discourse of the programme by virtue of his formal position within the structure of the text and narrative (the 'over-look'/Fiske as 'ideal viewer'). However, while Fiske's contribution is a powerful attempt at closure of the potential decodings of the programme, that power derives from two sources: both from his formal position within the text and from his social status (derived from a point outside the field of media discourse although necessarily realised within it) as an 'expert' on the relevant subject. It also depends on what he says, and the relation of the discourse within which he speaks to the discourses of the various sections of his audience.

This cuts both ways: if a member or section of the audience accepts the 'positionality' offered to him/her through Fiske's account of the problem then he/she does so as a result of both factors. Moreover (and this I think Heath and Skirrow would find it hard to account for), because both factors are relevant it is possible for the decoder to refuse or negotiate the 'position' offered through Fiske, despite his powerful structuring position in the text, if that interpellation is blocked or contradicted by the interpellations of other discourses in which the subject is involved, say as parent, truant or politician.

Taking up Neale's (1977) distinction between mode of address (textual characteristics) and ideological problematic we may attempt to reformulate the *Nationwide* model with reference to both dimensions. This is to argue for an articulation between the formal qualities of the text and the field of representations in and on which it works, and to pose the ideological field as the space in which signification operates. I would argue with Neale that:

> 'what marks the unity of an ideology is its problematic, the field and range of its representational possibilities (a field and range governed by the conjuncture) rather than any specific system of address.' (p. 18–19)

Our attempt in *Everyday Television: 'Nationwide'* to identify such a 'unity' at the level of problematic was couched descriptively. We attempted to construct a sentence: 'England is a country which is also a Nation in the modern world, with its own special heritage and problems' which would identify the particular 'combinatory' which constitutes *Nationwide*'s transformation and unification of elements from other ideological fields. Here again, Neale theorises the necessity of this process in his recognition that:

> 'an ideology . . . is never . . . a single, self-contained discourse or unit of themes, but is rather the product of other ideologies and discourses converging around and constituting those themes.' (p. 18)

The absence of this level of attention is apparent in the Heath and Skirrow piece. As O'Shea argues, what they end up doing, because they are not concerned with the programme's specific ideological formulation of the problems of the educational system, is to import through the back door the assumption that the programme is simply about 'truancy' as if this 'subject' existed unproblematically as a shared referent for all members of the audience rather than being, necessarily, constructed in its specific form through the process of signification.

Discourse Analysis
What is necessary is an attention to the level of 'referential potentiality' of signs and words – rather than the assumption of a model in which:

> 'words are thought of as labels applied to real or fictitious entities' (Henry, 1971, p. 77)

We must move to the level of discourse analysis:

> 'The emphasis upon referential potentialities of content words, as opposed to the notion of rigid connections between words and referents, moreover, paves the way for hypotheses concerning ways in which "referents" are established in the process of social interaction . . . discourse must be understood as a process of construction of social realities and imposition of these realities upon the other members of the communication process rather than as an exchange of information about externally defined referents.' (Carswell and Rommetreit, 1971, pp. 9–10)

This is to argue that contexts (both intra and extra discursive contexts) determine which referential potentialities will be activated in any particular case: the point being that referents are connected to discourses, not languages. In this sense there is no given referent for the word 'truancy', for example: referents are relative to discourses produced and interpreted in given conditions, not to 'dictionary meanings':

> 'Words, expressions and propositions have no meaning other than in their usage within a determinate discursive formation.' (Woods, 1977, p. 75)

The analysis must aim to lay bare the structural factors which determine the relative power of different discursive formations in the struggle over the necessary multi-accentuality of the sign – for it is in this struggle over the construction and interpretation of signs that meaning, (for instance the meaning of 'truancy') is produced. The crucial thing here is the 'insertion of texts into history via the way they are read' in specific socio-historical conditions, which in turn determine the relative power of different discursive formations. This is to recognise the determining effects of socio-historical conditions in the production of meaning:

'the relationship between the significations of an utterance and the socio-historical conditioning of that utterance's production is not of secondary importance but is constitutive of signification.' (Woods, p. 60)

Meaning must be understood as a production always within a determinate discursive formation; this is to abandon Saussure's equation of meaning with linguistic value, for his formulation:

'disguises the fact that within a single language community there can be distinct systems of linguistic value. That is to say, words and expressions (as signifiers) can change their meaning according to the ideological positions held by those who use them.' (Woods, p. 60)

This is to insist on the social production of meaning and the social location of subjectivity/ies – indeed it is to locate the production of subjectivity always within specific discursive formations.

Woods may provide us with a way of thinking the specificity of signification or discursive practice within ideology. He quotes Balibar, who argues that:

'if language [langue] is "indifferent" to class divisions and struggle, it does not follow that classes are indifferent to language. They utilise it, on the contrary, in a definite way in the field of their antagonism, notably in their class struggle.' (p. 60)

Woods goes on (and the argument can be extended to cover the specificity of the TV discourse):

'the indifference of language to class struggle characterises the "relative autonomy of the linguistic system", i.e. the set of phonological, morphological and syntactic structures, each with their own internal laws, constitutes the subject of linguistics. The non-indifference of classes to language, their specific utilisation of it in the production of discourses, is to be explained by the fact that every discursive process is inscribed in ideological class relations.' (p. 60)

It is this latter connection that is generally absent in the *Screen* position, e.g. in Heath and Skirrow, whose approach parallels that of the linguists whom Calvert criticises for failing to study:

'the status of language itself in society: its role in the class struggle, its ideological determination, etc. [This linguistics] contents itself with studying language [or signification] as a closed system, as one studies a mechanism . . . as a "neutral instrument" excluded from the field of social and political relations.' (quoted in Marina de Camargo, 1973, p. 10)

You cannot analyse signification without reference to the ideological context in which it appears; there is no way you can 'read' signification (e.g.

a TV programme) without knowing what the words mean in the specific cultural context.

The tendency towards an abstracted analysis, divorced from ideological and connotational context, concentrating on the text or 'product', is also criticised by Voloshinov. As Henriques and Sinha (1977) point out:

> 'For Voloshinov the proper object of study for a science of ideology is the living socially embedded utterance in context. It is the process of production and generation, not its abstracted and reified product, that provides the key to understanding the expression of ideology in consciousness and communication.' (p. 95)

This, then, is to situate signifying practice always within the field of ideology and class relations – as Henry (1971) puts it:

> 'The study of discursive processes, of discourses functioning in connection with other discourses, implies that we work with a corpus of messages and that we take into account the loci of speakers and addressees . . .
> The question is . . . how sentences . . . are interpreted in different ways depending on the loci at which they are assumed to be produced . . . ' (pp. 84–9)

Carswell and Rommetveit (op. cit., p. 5) make the proposition that we must 'expand our analysis of the utterance from its abstract syntactic form via its content, toward the patterns of communication in which it is embedded'. Substituting the terms of the argument, we can propose that we must expand our analysis of the TV text from its abstract signifying mechanisms (or mode of address) via its ideological themes towards the field of interdiscourse in which it is situated.

The Structure of Readings

Henry distinguishes between locus (in the social formation) and position (in discourse), arguing for the determination of position by locus:

> 'economic, institutional and ideological factors [determine] the locus occupied by an individual in the social structure. These factors constitute the conditions of production of the individual's discourses and the conditions of interpretation of those he receives. Through these conditions of production, the range and types of positions a given individual can adopt are determined . . . conditions of production and interpretation of discourses are tied to the different loci assigned to people by social structure.' (pp. 83–4)

Thus, the subject's position in the social formation structures his or her range of access to various discourses and ideological codes, and correspondingly different readings of programmes will be made by subjects 'inhabiting' these different discourses. What are important here are the relations of force of the competing discourses which attempt to 'interpellate' the subject: no one discourse or ideology can ever be assumed to have finally or fully dominated and enclosed an individual or social group. We must not assume that the dominant ideological meanings presented through television programmes have immediate and necessary effects on the audience. For some sections of the audience the codes and meanings of the programme will correspond more or less closely to those which they already inhabit in their various institutional, political, cultural and educational engagements, and for these sections of the audience the dominant readings encoded in the programme may well 'fit' and be accepted. For other sections of the audience the meanings and definitions encoded in a programme like *Nationwide* will jar to a greater or lesser extent with those produced by other institutions and discourses in which they are involved – trade unions or 'deviant' subcultures for example – and as a result the dominant meanings will be 'negotiated' or resisted. The crucial point to be explored is which sections of the audience have available to them what forms of 'resistance' to the dominant meanings articulated through television.

Texts, Readers and Subjects

The problem with much of *Screen's* work in this area has recently been identified by Willemen (1978) as the unjustifiable conflation of the reader of the text with the social subject. As he argues:

> 'There remains an unbridgeable gap between "real" readers/authors and "inscribed" ones, constructed and marked in and by the text. Real readers are subjects in history, living in social formations, rather than mere subjects of a single text. The two types of subject are not commensurate. But for the purposes of formalism, real readers are supposed to coincide with the constructed readers.' (p. 48)

Hardy, Johnston and Willemen (1976) provide us with the distinction between the 'inscribed reader of the text' and the 'social subject who is invited to take up this position'. They propose a model of 'interlocking subjectivities, caught up in a network of symbolic systems' in which the social subject:

> 'always exceeds the subject implied by the text because he/she is also placed by a heterogeneity of other cultural systems and is never coextensive with the subject placed by a single fragment (i.e. one film) of the overall cultural text . . .' (p. 5)

Nevertheless, the structure of discourse has its own specific effectivity, since:

> 'the social subject is also restricted by the positionality which the text offers it.'

The 'persons' involved are always already subjects within social practices in a determinate social formation; not simply subjects in 'the symbolic' in general, but in specific historical forms of sociality:

> 'this subject, at its most abstract and impersonal, is itself in history; the discourses . . . determining the terms of its play change according to the relations of force of competing discourses intersecting in the plane of the subject in history, the individual's location in ideology at a particular moment and place in the social formation.' (Willemen, 1978, pp. 63–4)

In his editorial to *Screen* v.18. n.3 Nowell-Smith rightly points to the particularity of Neale's approach, breaking, as it does, with the ahistorical and unspecified use of the category of the 'subject'. In his summary of Neale's position, Nowell-Smith points out that:

> '[Propaganda] . . . films require to be seen, politically, in terms of the positionality they provide for the socially located spectator.'

and this is 'on the one hand a question of textual relations proper, of mode of address' but also of 'the politico-historical conjuncture' because 'the binding of the spectator takes place' (or indeed fails to take place, we must add) 'not through formal mechanisms alone but through the way social institutions impose their effectivity at given moments across the text and also elsewhere.'

Woods' article on Pecheux provides us with the concept of interdiscourse in which the subject must be situated. He argues that:

> 'the constitution of subjects is always specific in respect of each subject . . . and this can be conceived of in terms of a single, original (and mythic) interpellation – the entry into language and the Symbolic – which constitutes a space wherein a complex of continually interpellated subject-forms inter-relate, each subject-form being a determinate formation of discursive processes. The discursive subject is therefore an interdiscourse, the product of the effects of discursive practices traversing the subject throughout its history.' (p. 75)

Alan O'Shea (1979) links this argument to Laclau's usage of the Althusserian notion of interpellation. For Laclau:

> 'the ideological field contains several "interlocking and antagonistic" discourses . . . any individual will be the "bearer and point of intersection" of a number of these discourses (in relations of dominance and subordination); and, as in the social formulation as a whole, within the individual these subjectivities will be articulated in a "relative coherence" . . .' (Laclau, 1977, p. 163)

Thus, O'Shea agrees, the individual is:

> 'the site of a particular nexus of the available ideological discourses, i.e.

> having a particular relationship with the ideological field, itself structured in dominance in a relation of dominant and dominated articulations. So just as there is a class struggle across the ideological level of the social formation, there will be a form of struggle within the individual's set of interpellations . . . which, like ideological discourses, can displace antagonism into difference in times of stability, but in crisis will be the site of several contradictory "hailings" one of which will have to prevail.' (p.4)

Thus, the 'subject' exists only as the articulation of the multiplicity of particular subjectivities borne by an individual (as legal subject, familial subject, etc.), and it is the nature of this differential and contradictory positioning within the field of ideological discourse which provides the theoretical basis for the differential reading of texts: the existence of differential positions in respect to the position preferred by the text. As Sylvia Harvey (1978) puts it:

> 'authors and readers can be studied in relation to those often different codes (processes for the production of meaning) for which they are the site of intersection. The encounter between reader and text is thus the complex sum of a number of different histories, the histories of the different codes for which author, text and readers are the site of intersection.' (p. 114)

In conclusion I want to argue that these concepts may now allow us to advance beyond two key problematics, one derived from *Screen*, one derived from Frank Parkin. The *Screen* one is that which Neale characterises as an 'abstract text-subject relationship', in which the subject is considered only in relation to one text, and that relation is contextualised only by the primary psychic processes of the mirror/Oedipal phases, etc., which the present relation replays, or which provide the foundation of the present relation.

Parkin provides us with a sociologically informed critique of the abstract relation outlined above but into which demographic/sociological factors, age, sex, race, class position, etc., are introduced as directly determining subject responses. These factors, however, have often been presented as objective correlates/determinants, for instance, of differential decoding positions without specification of how they intervene in the process of communication.

Now while I would insist on the centrality of the question of structural determination and the articulation of signifying practices with the field of ideology, as it is structured by the 'real', nevertheless we must accept with Willis (1978) that:

> 'It cannot be assumed that cultural forms are determined in some way as an automatic reflex by macro determinations such as class location, region, and educational background. Certainly these variables are

important and cannot be overlooked, but *how* do they impinge on behaviour, speech and attitude? We need to understand how structures become sources of meaning and determinations on behaviour in the cultural milieu at its own level . . . the macro determinants need to pass through the cultural milieu to reproduce themselves at all.' (p. 171)

We need to construct a model in which the social subject is always seen as interpellated by a number of discourses, some of which are in parallel and reinforce each other, some of which are contradictory and block or inflect the successful interpellation of the subject by other discourses. Positively or negatively, other discourses are always involved in the relation of text and subject, although their action is simply more visible when it is a negative and contradictory rather than a positive and reinforcing effect.

We cannot consider the single, hypostatised text-subject relation in isolation from other discourses. Neither should we 'read in' sociological/demographic factors as directly affecting the communication process: these factors can only have effect *through* the action of the discourses in which they are articulated. The notions of interdiscourse, and of multiple and contradictory interpellations of the subject, open up the space between text and subject. We no longer assume that the subject is effectively bound by any particular interpellation, and thus provide the theoretical space for the subject to be in some other relation to the signifying chain from that of 'regulated process'.

Crucially, we are led to pose the relation of text and subject as an empirical question to be investigated, rather that as an *priori* question to be deduced from a theory of the ideal spectator 'inscribed' in the text. It may be, as MacCabe (1976, p.25) emphasises, that analysing a film within a determinate social moment in its relation to its audience 'has nothing to do with the counting of heads' but this is a point of methodological adequacy, not of theoretical principle. The relation of an audience to the ideological operations of television remains in principle an empirical question: the challenge is the attempt to develop appropriate methods of empirical investigation of that relation.

Afterword

This is not a conclusion, as the very nature of the research project on which this monograph is based precludes any such finality. The project has been conducted as a preliminary investigation, at the empirical level, of the potential of the 'encoding/decoding' model as applied to the TV audience.

In its original form the research was designed to cover both a wider range of televisual material and a wider range of groups, as well as a more intensive examination of group and individual readings. However, this plan was premised on a level of funding (for two full time researchers) which was not available. I can only express my gratitude here to the BFI for providing the funds which enabled the project to take place at all – though, of necessity, it was scaled down to the level of one, part-time, researcher, plus a considerable amount of voluntary help.

I have been able to do no more than to indicate some of the ways in which social position and (sub)cultural frameworks may be related to individual readings. To claim more than that, on the basis of such a small sample, would be misleading. Similarly, I would claim only to have shown the viability of an approach which treats the audience as a set of cultural groupings rather than as a mass of individuals or as a set of rigid socio-demographic categories. Clearly, more work needs to be done on the relation between group and individual readings.

However, while these are important considerations for further research, the limitation of which I am most conscious, and which I would see as the priority for further work in this area, lies at the level of methodology. Here the question is one of developing methods of analysis which would allow us to go beyond the descriptive form in which the material is handled here.

The material from the groups is presented at such length here precisely because of the absence of any adequate method which would enable us to formalise and condense the particular responses into consistent linguistic and/or ideological categories. The difficulty here is obviously that of the complex and overdetermined nature of the relation between linguistic (or discursive) and ideological features. In the absence of any method capable of satisfactorily meeting these difficulties, it seemed more useful to present the material in a descriptive format, in the hope that it would then be more 'open' to the reader's own hypotheses and interpretations where mine seem inadequate. It is in this area, of the relations between linguistic, discursive forms and particular ideological positions and frameworks, that this work most needs to be developed.*

*For some attempts in this direction see Kress and Trew (1979) and Fowler *et al.* (1979).

As the reader will no doubt have noted, the theoretical framework which informs this work is presented in relation to quite different theoretical debates in different chapters – at one point in relation to work in mainstream sociology, at another in relation to psychoanalytic film theory. Over the project's long period of gestation it has been developed through and against these varied theoretical protagonists and it seemed best to leave the marks of these debates visible, so as to make it clearer where the perspective employed has developed from. If it then seems that I have taken us down some wrong turnings, it may be possible for the reader to more easily find his or her way back, or onwards, out of the swamp.

Bibliography

Althusser, L. (1971), 'Ideology and Ideological State Apparatuses' in *Lenin and Philosophy*, NLB, London.

Armistead, N. (ed.) (1974), *Reconstructing Social Psychology*, Penguin, Harmondsworth.

Bandura, B. (1961), 'Identification as a process of social learning' in *Journal of Abnormal and Social Psychology*, Vol. 63, No. 2.

Berelson, B. (1952), *Content Analysis in Communication Research*, Free Press, Glencoe, Ill.

Berkowitz, L. (1962), 'Violence and the mass media', in Paris Stanford Studies in Communication, Institute for Communication Research.

Bernstein, B. (1971), *Class, Codes and Control*, Paladin, London.

Beynon, H. (1973), *Working for Ford*, Penguin, Harmondsworth.

Blumler, J. and Ewbank, A. (1969), 'Trade Unionists, the media and unofficial strikes', in *British Journal of Industrial Relations*, 1969.

Bulmer, M. (ed.) (1977), *Social Research Methods*, Macmillan, London.

de Camargo, M. (1973), 'Ideological Analysis of Media Messages', CCCS mimeo, University of Birmingham.

Carswell, E. and Rommetveit, R. (eds.) (1971), *Social Contexts of Messages*, Academic Press, London.

Caughie, J. (1977), 'The world of television', in *Edinburgh '77 Magazine*.

Connell, I. (1978), 'The reception of television science', Primary Communications Research Centre, University of Leicester.

Connell, I. (1979), 'Ideology/Discourse/Institution', *Screen*, Vol. 19, No. 4.

Connerton, P. (ed.) (1976), *Critical Sociology*, Penguin, Harmondsworth.

Coward, R. (1977), 'Class, Culture and Social Formation', *Screen*, Vol. 18, No. 1.

Counihan, M. (1972), 'Orthodoxy, Revisionism and Guerilla Warfare in Mass Communications Research', CCCS mimeo, University of Birmingham.

Critcher, C. (1978), 'Structures, Cultures and Biographies', in Hall and Jefferson (1978).

Curran, J. *et al.* (eds.) (1977), *Mass Communications and Society*, Arnold, London.

Deutscher, I. (1977), 'Asking Questions (and listening to answers)', in Bulmer (ed.) (1977).

Downing, T. (1974), *Some Aspects of the Presentation of Industrial Relations and Race Relations in the British Media*, Ph.D. thesis, London School of Economics.

Dyer, R. (1977), '*Victim:* hermeneutic project', *Film Form*, Autumn 1977.

Eco, U. (1972), 'Towards a semiotic enquiry into the TV message' in *Working Papers in Cultural Studies*, No. 3, CCCS, University of Birmingham.

Elliott, P. (1972), *The making of a television series*, Constable, London.

Elliott, P. (1973), 'Uses and gratifications: a critique and a sociological alternative', Centre for Mass Communications Research, University of Leicester.

Ellis, J. (1977), 'The institution of the cinema', in *Edinburgh '77 Magazine*.

Fowler, R. *et al.* (1979), *Language and Control,* RKP, London.
Garnham, N. (1973), *Structures of Television,* BFI TV Monograph No. 1.
Gerbner, G. (1964), 'Ideological Perspectives in News Reporting', in *Journalism Quarterly,* Vol. 41, No. 4.
Giglioli, P. (1972), *Language and Social Context,* Penguin, Harmondsworth.
Grealy, J. (1977), 'Notes on Popular Culture', *Screen Education,* No. 22.
Hall, S. (1973), 'Encoding and Decoding the TV message', CCCS mimeo, University of Birmingham.
Hall, S. (1974), 'Deviancy, Politics and the Media', in Rock and McIntosh (eds.) (1974).
Hall, S. and Jefferson, T. (1978), *Resistance through Ritual,* Hutchinson, London.
Halloran, J. (ed.) (1970a), *The Effects of Television,* Panther, London.
Halloran, J. *et al.* (1970b), *Demonstrations and Communications,* Penguin, Harmondsworth.
Halloran, J. (1975), 'Understanding Television', *Screen Education,* No. 14.
Hardy, P. *et al.* (1976), 'Introduction', to *Edinburgh '76 Magazine.*
Hartmann, P. and Husband, C. (1972), 'Mass Media and Racial Conflict' in McQuail (ed.) (1972).
Harvey, S. (1978), *May '68 and Film Culture,* BFI, London.
Heath, S. (1976), 'Screen Images, Film Memory', in *Edinburgh '76 Magazine.*
Heath, S. and Skirrow, G. (1977), 'Television: a world in action', *Screen,* Vol. 18, No. 2.
Henriques, J. and Sinha, C. (1977), 'Language and Revolution', in *Ideology and Consciousness,* No. 1.
Henry, P. (1971), 'On processing message referents in context', in Carswell and Rommetveit (eds.) (1971).
Hirst, P. (1976), 'Althusser's theory of ideology', in *Economy and Society,* November 1976.
Katz, E. (1959), 'Mass communications research and the study of popular culture', *Studies in Public Communication,* Vol. 2.
Katz, E. and Lazarsfeld, P. (1955), *Personal Influence,* Free Press, Glencoe, Ill.
Klapper, J. (1960), *The Effects of Mass Communication,* Free Press, Glencoe, Ill.
Kress, G. and Trew, T. (1979), 'Ideological transformation of discourse', *Sociological Review,* May 1979.
Kumar, K. (1977), 'Holding the Middle Ground', in Curran (ed.) (1977).
Laclau, E. (1977), *Politics and Ideology in Marxist Theory,* NLB, London.
Lazarsfeld, P. and Rosenberg, M. (eds.) (1955), *The Language of Social Research,* Free Press, New York.
Linné, O. and Marossi, K. (1976), *Understanding Television,* Danish Radio Research Report.
Mann, M. (1973), *Consciousness and Action among the Western Working Class,* Macmillan, London.
MacCabe, C. (1976), 'Realism and Pleasure', in *Screen,* Vol. 17, No. 3.
McQuail, D. (ed.) (1972), *Sociology of Mass Communication,* Penguin, Harmondsworth.
Merton, R. (1946), *Mass Persuasion,* Free Press, New York.
Merton, R. and Kendall, P. (1955), 'The Focussed Interview', in Lazarsfeld and Rosenberg (1955).
Merton, R. (ed.) (1959), *Sociology Today,* Free Press, New York.

Miliband, R. and Saville, J. (eds.) (1973), *Socialist Register,* Merlin, London.
Mills, A. and Lowe, A. (1978), 'Screen and Realism', CCCS mimeo, University of Birmingham.
Mills, C. W. (1939), 'Language, Logic and Culture', in *Power, Politics and People,* OUP, London and New York.
Moorhouse, H. and Chaimberlain, C. (1974a), 'Lower class attitudes to the British Political System', in *Sociological Review,* Vol. 22, No. 4.
Moorhouse, H. and Chaimberlain, C. (1974b), 'Lower class attitudes to property', in *Sociology,* Vol. 8. No. 3.
Morley, D. (1976), 'Industrial Conflict and the Mass Media', *Sociological Review,* May 1976.
Murdock, G. and Golding, P. (1973), 'For a political economy of mass communications', in Miliband and Saville (eds.) (1973).
Murdock, G. (1974), 'Mass communication and the construction of meaning', in Armistead (ed.) (1974).
Neale, S. (1977), 'Propaganda', in *Screen,* Vol. 18, No. 3.
Nicholls, T. and Armstrong, P. (1976), *Workers Divided,* Fontana, London.
Nordenstreng, K. (1972), 'Policy for news transmission', in McQuail (ed.) (1972).
O'Shea, A. (1979), 'Laclau on Interpellation', CCCS mimeo, University of Birmingham.
Parkin, F. (1973), *Class Inequality and Political Order,* Paladin, London.
Piepe, A. *et al.* (1978), *Mass Media and Cultural Relationships,* Saxon House, London.
Pollock, F. (1976), 'Empirical research into public opinion', in Connerton (ed.) (1976).
Poulantzas, N. (1973), *Political Power and Social Classes,* NLB, London.
Riley, J. and Riley, M. (1959), 'Mass communications and the social system', in Merton (ed.) (1959).
Rock, P. and McIntosh, M. (1974), *Deviance and Social Control,* Tavistock, London.
Rosen, H. (1972), *Language and Class,* Falling Wall Press, London.
Stevens, T. (1978), 'Reading the realist film', *Screen Education,* No. 26.
Thompson, E. P. (1978), *The Poverty of Theory,* Merlin, London.
Thompson, J. O. (1978), 'A nation wooed', *Screen Education,* No. 29.
Voloshinov, V. (1973), *Marxism and the Philosophy of Language,* Academic Press, New York.
Willemen, P. (1978), 'Notes on subjectivity', *Screen,* Vol. 19, No. 1.
Willis, P. (1978), *Learning to Labour,* Saxon House, London.
Woods, R. (1977) 'Discourse analysis', in *Ideology and Consciousness,* No. 2.
Woolfson, C. (1976), 'The semiotics of workers' speech', in *Working Papers in Cultural Studies,* No. 9, CCCS, University of Birmingham.
Wright, C. R. (1960), 'Functional analysis and mass communication', *Public Opinion Quarterly,* 24.

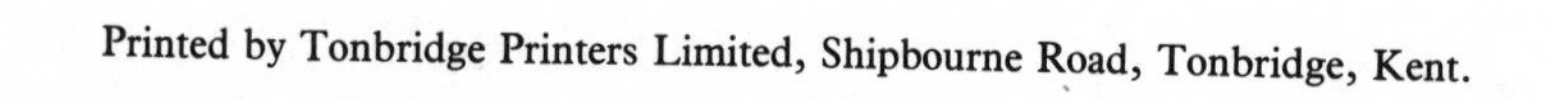

Printed by Tonbridge Printers Limited, Shipbourne Road, Tonbridge, Kent.